Information Circular 9507

Reducing Low Back Pain and Disability in Mining

By Sean Gallagher, Ph.D., CPE

DEPARTMENT OF HEALTH AND HUMAN SERVICES
Centers for Disease Control and Prevention
National Institute for Occupational Safety and Health
Pittsburgh Research Laboratory
Pittsburgh, PA

September 2008

This document is in the public domain and may be freely copied or reprinted.

Disclaimer

Mention of any company or product does not constitute endorsement by the National Institute for Occupational Safety and Health (NIOSH). In addition, citations to Web sites external to NIOSH do not constitute NIOSH endorsement of the sponsoring organizations or their programs or products. Furthermore, NIOSH is not responsible for the content of these Web sites.

Ordering Information

To receive documents or other information about occupational safety and health topics, contact NIOSH at

> Telephone: **1–800–CDC–INFO** (1–800–232–4636)
> TTY: 1–888–232–6348
> e-mail: cdcinfo@cdc.gov
>
> or visit the NIOSH Web site at **www.cdc.gov/niosh**.

For a monthly update on news at NIOSH, subscribe to NIOSH *eNews* by visiting **www.cdc.gov/niosh/eNews**.

DHHS (NIOSH) Publication No. 2008–135

September 2008

SAFER • HEALTHIER • PEOPLE™

CONTENTS

Page

Glossary vii
Acknowledgments viii
Executive summary 1
Introduction 2
I. Understanding low back pain 3
    The nature of low back pain 3
    Causes of low back pain 6
    Sources of low back pain (pain generators) 10
II. Low back disability in mining 14
    Reportable back injuries in mining (1996–2005) 14
    Unique demands of the mining environment 19
    Summary 25
III. Primary prevention of low back pain in mining 26
    Establishing a proactive process for preventing low back pain 26
    Facilities design and layout 29
    Use of mechanical-assist devices 34
    Application of mechanical-assist devices in mining 35
    The design of lifting tasks 42
    Improving the design of lifting tasks 46
    Principles of seating and control of whole-body vibration 51
    Psychosocial factors, work stress, and low back pain 56
IV. Secondary prevention: reducing disability 58
    Workplace design 58
    Proactive return-to-work program 59
    Communication 60
    Management commitment 60
    Wellness behaviors during low back pain episodes 61
V. Summary and guidelines 63
VI. Directions for future research 65
References 67
Appendix.—Tables used for calculation of the revised NIOSH lifting equation 76

ILLUSTRATIONS

1. Regions of the human spine 6
2. A motion segment of the lumbar spine 7
3. Top and side views of lumbar vertebra 7
4. Vertebral endplates are located on the vertebral body adjacent to the discs 8
5. Muscles of the low back 9
6. Ligaments of the lumbar spine 9
7. The endplate is the first structure to become damaged when subjected to a high or
    repeated load 10
8. Fissures (tears) in the disc 11

ILLUSTRATIONS - Continued

Page

9. Adopting a full stooping posture 13
10. Back injury rates per 100 FTEs over all mining commodities and contractors (1996–2005) ... 15
11. Back injury rates per 100 FTEs, surface versus underground (1996–2005) 15
12. Back injury rates by commodity and mine type (surface versus underground) aggregated from 1996 to 2005 16
13. Percentage of back injuries by mine worker activity (1996–2005) 17
14. The seven leading job titles in terms of the percentage of back injuries reported during 1996–2005 18
15. One-lift maximum lifting capacity for males in various postures 22
16. One-lift maximum lifting capacity for females in various postures 22
17. Comparison of strength measures for coal miners working in confined vertical space to an industrial population 23
18. Lumbar moments, an indicator of load imposed on the low back, are increased as vertical workspace becomes more confined 24
19. The risk management cycle 26
20. Comparison of present and proposed methods for installing roof screen in an underground coal mine 33
21. Hoists can be used at many locations on the surface and underground to lift heavy items 35
22. Using an air bag to lift a conveyor belt 36
23. Mobile manipulator 36
24. Using forklift and pallet for easy transfer of loads 37
25. Clevis pin removal 37
26. Counterweight installed to facilitate opening of vault 38
27. Lift stands can have many uses in shop areas and can greatly facilitate load transfer 38
28. Attaching a simple handle to wheel chocks greatly reduces excessive bending to place or remove the chocks 39
29. Using steel bar to open hopper gates *(left)*; gates opened via mechanical assist *(right)* 39
30. Installing railcar shaker involves heavy load in flexed posture *(left)*; a simple device can be used to install shaker with a minimum of effort *(right)* 40
31. Well-organized trays *(left)* can be efficiently transported underground using scoop-mounted forks *(right)* 40
32. Use of a winch to facilitate loading roof screen 41
33. Cart that rides on conveyor handrails reduces the demands of transporting supplies down the longwall 41
34. Examples of holding a crib block 43
35. In this example, the back muscles have a 2-in lever arm compared to the 20-in lever arm for the load 44
36. Since weight of the load is multiplied by the distance away, a lighter object lifted at a great distance from the body can actually create more spine stress than a heavy one lifted close to the body 45
37. Suggested dimensions for design of lumbar support 52
38. Seating space envelopes for 95th-percentile miners in 42-in and 22-in compartments 55
39. Prototype seat design for underground mining equipment 56

TABLES

Page

1. Work-related physical risk factors for low back disorders 4
A-1. Frequency multiplier table 76
A-2. Coupling multiplier table 76

ACRONYMS AND ABBREVIATIONS USED IN THIS REPORT

ANSI	American National Standards Institute
ATRS	automated temporary roof support
FTE	full-time equivalent
HFES	Human Factors and Ergonomics Society
ISO	International Organization for Standardization
LBP	low back pain
LI	lifting index
MSHA	Mine Safety and Health Administration
NIOSH	National Institute for Occupational Safety and Health
PPE	personal protective equipment
RNLE	Revised NIOSH Lifting Equation
RWL	recommended weight limit
WBV	whole-body vibration

UNIT OF MEASURE ABBREVIATIONS USED IN THIS REPORT

cm	centimeter
ft	foot
ft-lb	foot-pound
hr	hour
Hz	hertz
in	inch
in-lb	inch-pound
kg	kilogram
lb	pound
min	minute
N	newton

GLOSSARY

Disc degeneration: A process through which the intervertebral disc begins to break down, leading to fissures and tears in the disc. These fissures or tears can become painful if they extend to (or occur in) the outer portion of the disc.

Fatigue failure: A process of material failure that results from repeated loading. Typically, the process begins with development of a small crack in the material due to an applied stress. Further loading expands the crack until the material fails.

Fracture: A break in bone or cartilage resulting from exposure to high forces and/or repeated loading.

Inflammation: A basic way in which the body reacts to infection, irritation, or other injury. The key features are redness, warmth, swelling, and pain.

Intervertebral disc: A disc-shaped piece of specialized tissue that separates the bones of the spinal column.

Low back disability: Lost time from work resulting from low back pain.

Low back pain (LBP): The subjective perception of pain in the low back, buttocks, or legs.

Lumbar vertebrae: Refers to the five lumbar bones (vertebrae) of the spine, situated in the lower back region between the thoracic vertebrae and the sacral vertebrae in the spinal column. These vertebrae are subjected to high stress and are a frequent site of injury.

Moment: The tendency of a force to cause rotation about a point or axis. A moment represents the combined effect of a *force* applied at a *distance* from a point.

Ultimate compressive strength: The force that a material can resist without being crushed, under pressure, in a single loading cycle.

Vertebral endplate: A thin layer of cartilage that connects the vertebrae and the intervertebral discs of the spine. Endplates are the weakest link in the spine and the first structure to get damaged when the spine is placed under load.

Whole-body vibration (WBV): Mechanical energy oscillation transmitted to the whole body, typically during operation of mobile equipment or vehicles. Exposure to WBV is associated with increased incidence of low back pain.

ACKNOWLEDGMENTS

The author wishes to gratefully acknowledge Dr. Stover H. Snook for providing material used and referenced in this report. The author would also like to acknowledge technical review commentary received from Dr. Tom Waters (NIOSH), Dr. Robert Dick (NIOSH), Dr. Patrick Dempsey (NIOSH), Dr. David Gilkey (Colorado State University), Dr. G. T. Lineberry (University of Kentucky), Dr. Robin Burgess-Limerick (University of Queensland, Australia), Mr. Butch Demich (Rox Coal Co.), Mr. Steve Ziats (United Mine Workers of America), and Mr. Jon Montgomery (Mine Safety and Health Administration). Ms. Lisa Steiner (NIOSH) and Dr. Janet Torma-Krajewski (NIOSH) provided comments on earlier drafts that were beneficial in the development of this report. Photo credits are due Dr. Burgess-Limerick, Ms. Steiner, and Dr. Torma-Krajewski.

REDUCING LOW BACK PAIN AND DISABILITY IN MINING

By Sean Gallagher, Ph.D., CPE[1]

EXECUTIVE SUMMARY

This report was written in an effort to provide better control measures for low back pain (LBP) and low back disability in the mining industry. There are numerous factors associated with development of LBP, many of which can be effectively controlled while some cannot. Better job design is promoted as the best method of reducing cases of LBP and can also reduce the disability (i.e., lost time from work) associated with LBP when it happens.

The report draws attention to what is currently known about LBP, what the causes are thought to be, and discusses recent back injury trends in the mining industry. Research describing unique physical demands in mining, such as the capabilities and limitations of working in awkward postures, is also presented. Methods that can be used to prevent initial LBP episodes are provided, including facilities design and layout for materials and supplies, use of mechanical-assist devices, improved design of lifting tasks, and better seat design. Methods of reducing the disability associated with LBP (including workplace design, proactive return-to-work efforts, communication, and management commitment) are also discussed.

The report concludes that control of LBP and disability in mining requires a comprehensive approach to limit the repetitive loading that can occur on the low back due to manual materials-handling tasks and exposure to whole-body vibration (WBV). Specific recommendations include the following:

- Successful LBP prevention efforts require a proactive program that has strong management commitment and incorporates employee involvement.
- More efficient supply handling systems and use of mechanical-assist devices can greatly reduce exposure to hazardous lifting tasks.
- Lifting tasks should be designed to minimize low back stress. Tools to evaluate and redesign lifting tasks are presented.
- Improved seat design can reduce exposure to WBV and improve posture, leading to reduced LBP risk.
- The disability associated with LBP can be reduced. Getting the worker back on the job as quickly as possible is in the interest of everyone involved.

---

[1]Senior research scientist, Pittsburgh Research Laboratory, National Institute for Occupational Safety and Health, Pittsburgh, PA.

INTRODUCTION

Low back pain (LBP) and the disability resulting from LBP continue to plague the mining industry. Historically, the mining industry has been among the highest of industries in terms of LBP worker compensation claims [Klein et al. 1984], back injury rates [Sestito et al. 2004], and exposure to risk factors for back and other musculoskeletal disorders [Winn and Biersner 1992]. The high incidence of LBP among miners is believed to be due (at least in part) to high exposure to awkward postures, heavy manual work, and exposure to whole-body vibration (WBV) in the mining work environment [Brinckmann et al. 1998]. As a result of these risk factor exposures, LBP has consistently been the leading cause of occupational disability in the mining industry.

LBP prevention efforts in mining (and other industries) have often been defined by the well-known slogan: "Lift with your legs, not with your back." However, it is now clear that the challenge of preventing LBP and disability cannot be addressed so simply. The slogan implies that total ownership for LBP prevention lies entirely with the worker performing a lifting task. However, closer inspection reveals that lifting demands experienced by the worker may be the result of the workplace environment, job demands, and decisions made by others (e.g.,, by material suppliers or those deciding how materials will be stored). Furthermore, research has shown that the most effective efforts to control LBP and disability are those that are comprehensive in nature and that involve individuals at all levels of the organization (and often individuals *outside* the organization) [Snook 2006].

This report seeks to provide managers, supervisors, and safety and health professionals with a greater understanding of LBP and low back disability (work time lost due to LBP). The report will attempt to improve the reader's understanding of the many factors that influence LBP, provide the latest research information related to its causes, and describe methods that have proven to be effective in preventing LBP and disability.

I. UNDERSTANDING LOW BACK PAIN

The Nature of Low Back Pain

Developing a successful strategy for the control of LBP requires an understanding of the nature of the pain and factors that may influence its development. LBP is a complex phenomenon. While there is still a great deal that is not understood about back pain, much has been learned during the past decades. The sections that follow provide some basic information regarding what is currently known about LBP.

What Is Low Back Pain?

LBP has been defined as the subjective perception of pain in the low back, buttocks, or legs [Snook 2006]. Numbness and/or pain that radiates to the legs is commonly known as *sciatica*. LBP is a very common affliction. Approximately two-thirds of the U.S. adult population suffers from LBP at some time in their lives [Deyo and Weinstein 2001]. Pain is classified by health professionals as minimal, slight, moderate, or severe. Minimal pain constitutes an annoyance, but does not affect activity. Slight pain may have small effects on physical activity, while moderate pain may cause a marked handicap in activity. Severe pain may result in a severe handicap or may preclude performance of an activity altogether.

Why Is Low Back Pain Hard to Diagnose?

Since there is no objective measure of pain, subjective reports must be relied upon (pain drawings, pain scales, etc.). Due in part to this problem, about 85% of people with LBP are not given precise reason for their pain [Deyo and Weinstein 2001; Nachemson and Vingård 2000]. The most common diagnosis for most people is "nonspecific" LBP. As will be discussed in later sections, more is being learned about pain generators and pathways believed to lead to LBP. However, specific diagnoses as to the true cause of pain remain rare and complex at present.

How Often Does Low Back Pain Recur?

The recurrence rate for LBP is very high. A recurrence range of 40%–80% is commonly reported [Wasiak et al. 2003]. In fact, studies have often found that a history of LBP is the best predictor for future LBP [Videman and Battié 1996; Carter and Birrell 2000; Waddell and Burton 2000]. The high recurrence rate for LBP means that it is very important to consider secondary prevention (prevention of LBP recurrence) in efforts to control costs and disability.

What Is the Relationship Between Low Back Pain and Disability?

Low back disability (time lost from work) is obviously related to LBP, but includes many personal, social, and economic factors that can affect disability [Snook 2006]. Examples of these variables include job dissatisfaction, personality differences, supervisor conflicts, domestic problems, compensation laws, management policies and practices, unemployment rates, and litigation. The relationship between LBP and low back disability is not always straightforward. Some people experience high amounts of LBP, but can still work. Others can become disabled with only small amounts of LBP. The degree to which a worker is disabled is often related to the physical demands of the worker's job. For example, a worker in the mine office may be able to perform his or her duties with LBP, but the same amount of pain may make work unbearable to a coal miner who works in a low- or midseam mine. Furthermore, the physical stresses of work can sometimes exacerbate a mild degree of LBP to unbearable levels.

What Are the Occupational Risk Factors for Low Back Pain?

A risk factor is a variable associated with an increased risk of disease, disorder, or infection. Risk factors are not necessarily the cause of a disease or disorder. Risk factors are obtained by comparing the risk of those exposed to the potential risk factor to those not exposed. The term "risk factor" was first coined by heart researcher Thomas R. Dawber, M.D., in a landmark scientific paper in 1961 in which he attributed heart disease to specific conditions such as blood pressure, cholesterol, and smoking [Kannel et al. 1961].

A number of risk factors for LBP have been identified. Several of these are of interest in terms of exposures present in the mining environment. As shown in Table 1, a recent summary of the scientific literature performed by the National Research Council [NRC 2001] found that four work-related risk factors show consistent and positive associations with the occurrence of back disorders. These risk factors include manual materials handling, frequent bending and twisting, heavy physical load, and exposure to WBV. These results echo the findings of a comprehensive review by the National Institute for Occupational Safety and Health (NIOSH), which found strong evidence for a causal relationship between LBP and lifting/forceful movements and exposure to WBV and evidence for a causal relationship between awkward postures and heavy physical work and LBP [Bernard 1997].

Table 1.—Work-related physical risk factors for low back disorders [NRC 2001]
(n = number of studies)

Risk factors	Studies showing no association with back disorders		Studies showing positive association with back disorders	
	n	Range of risk estimates	n	Range of risk estimates
Manual material handling	4	0.90–1.45	24	1.12–3.54
Frequent bending and twisting	2	1.08–1.30	15	1.29–8.09
Heavy physical load	0	—	8	1.54–3.71
Whole-body vibration	1	1.10	16	1.26–9.00

It should be apparent that all of the physical risk factors listed above have relevance to work in the mining industry. For example, heavy physical work has been defined as work that requires high energy demands or some measure of physical strength [Bernard 1997]. The mining industry is filled with tasks that require high physical demands. Furthermore, manual lifting and handling of heavy supplies and other material are still commonplace in mining. Underground mining often presents workers with unique environmental restrictions that force the use of awkward working postures involving frequent and severe bending and twisting activities. However, bending and twisting are certainly not unknown in surface mining operations either. Finally, exposure to WBV is common for mine workers riding in or on equipment at both surface and underground settings. Exposures to WBV in mining can be exacerbated by driving trucks on rough haul roads at surface mines and by the inability to put effective vehicle suspension systems on low-seam mobile equipment.

Personal Factors

There are a number of personal factors that seem to influence risk of LBP [NRC 2001; Snook 2006]. These include factors such as age, physical strength, prior injury, level of individual fitness and conditioning, pain intolerance, and certain psychological disorders. Some of these issues (such as the effects of age and prior injury) are complex and are clearly not factors that can be completely controlled at the present time. However, proper design of jobs can do much to accommodate individuals of different ages and capabilities and can also influence certain psychological factors that have shown associations with LBP and low back disability.

Does the Social Environment at Work Affect Low Back Pain?

Psychosocial factors, such as job dissatisfaction, poor relationship with immediate supervisors, perceived inadequacy of income, lack of control over one's job, and unpleasant work environments seem to have an impact on LBP and/or reporting of LBP [Bongers et al. 1993; NRC 2001; Bernard 1997; Snook 2006]. One of the consistent findings related to workplace psychosocial stressors is that low job satisfaction is associated with LBP [Westrin 1970; Magora 1973; Svennson and Andersson 1983; Bergenudd and Nilsson 1988; Hoogendoorn et al. 2000]. Monotony at work is also usually associated with higher levels of LBP and worker compensation claims. Monotony was found to have a direct relationship to LBP in a study by Svennson and Andersson [1983], while a study by Bergqvist-Ullman and Larsson [1977] found that monotonous jobs were also associated with increased lost time. Other workplace factors that have shown a relationship to LBP include low social support at work, high perceived stress, and perceived ability to return to work [Hoogendoorn et al. 2000, 2001; NRC 2001]. Recent studies have indicated that the combination of workplace stress and physical work demands increase rates of LBP well above the rates seen due to physical demands alone [Devereux et al. 2004; Waters et al. 2007]. One of these studies found that each factor alone doubled the risk of LBP, but if both factors were present the risk was four times higher [Waters et al. 2007]. Furthermore, reviews of research indicate that psychosocial and psychological issues may be particularly important in the transition from LBP to disability [Ferguson and Marras 1997; Linton 2000]. Improving social aspects of the work environment can reduce the psychosocial factors that lead to LBP and low back disability.

Summary

LBP is a complex problem that involves both factors that cannot currently be controlled (such as certain effects of age) and factors that can be changed or influenced (such as the design of jobs, equipment, and work culture). Control of LBP in mining should focus on both reducing exposure to physical risk factors to prevent initial onset of LBP and designing work to better accommodate workers with LBP. Fortunately, many of the techniques covered in this report can help achieve both objectives. Before discussing these control techniques, the causes of LBP and recent back injury trends in the mining industry will be examined.

Causes of Low Back Pain

In past years, relatively little was known about specific causes of LBP. However, recent research has improved our understanding of some of the causes of back pain. While there are many possible sources of LBP, some seem to be more highly associated with chronic LBP (and are responsible for a large portion of low back disability). An understanding of these mechanisms requires a brief look at the anatomy of the low back and a discussion of the injury pathways leading to LBP.

Anatomy of the Lumbar Spine

Figure 1 shows the entire human spine and its major regions. The lumbar (low back) region of the spine is located directly below the thoracic region and directly above the sacrum. The lumbar vertebrae are most frequently involved in LBP because these vertebrae carry the greatest amount of body weight and are subject to the largest forces and stresses in the spine.

Figure 2 illustrates what is known as a *motion segment*, which consists of two vertebrae and a spinal disc (called an intervertebral disc). The disc serves an important function in that it provide a means of separating individual spinal vertebrae and, being deformable, allows flexibility of movement to the spinal column. If the spine were just a stack of bones with no intervening discs, the spine could bear weight but would not be able to provide the rocking motions that allow forward and side bending of the spine, which would greatly limit the types of tasks that humans could perform. Thus, the motion segment does what it says—it is a segment of the spine that allows motion to occur. Each individual motion segment provides only a relatively small amount of movement (a few degrees in any direction). However, when one adds up these small movements across all of the motion segments of the spine, the result is tremendous flexibility in forward bending, side bending, and twisting motions.

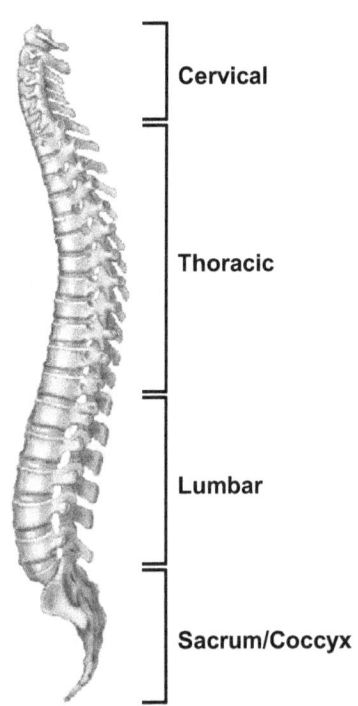

Figure 1.—Regions of the human spine.

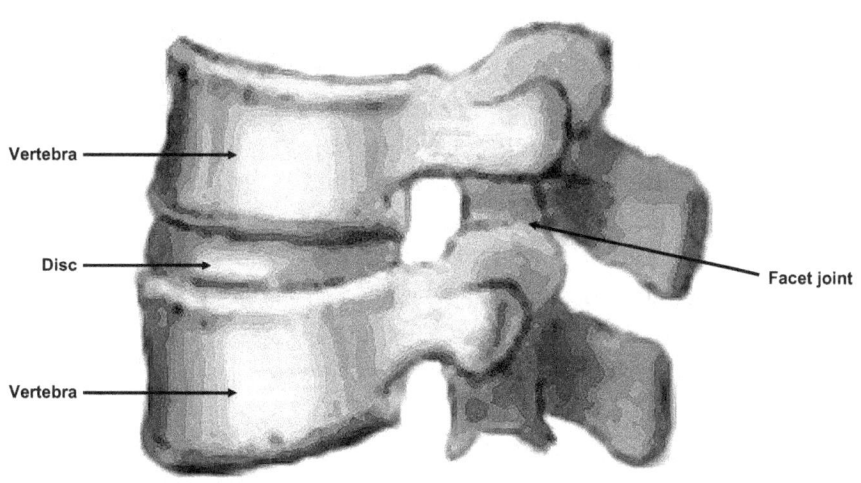

Figure 2.—A motion segment of the lumbar spine.

A closer look at the components of the motion segment will now be taken, as some of these components seem to play important roles in the development of LBP.

The Lumbar Vertebrae

Lumbar vertebrae are irregularly shaped bones that consist of three major functional elements: the vertebral body, the pedicles, and what are called posterior elements (Figure 3). The vertebral body is a short section of a cylinder, but when viewed from above, reveals a somewhat kidney-shaped appearance. The "walls" of the cylinder consist of dense bone (cortical bone) that provides a strong exterior structure. Within this ring of cortical bone there is a less dense type of bone called cancellous bone (which, while made of bone, looks somewhat "spongelike").

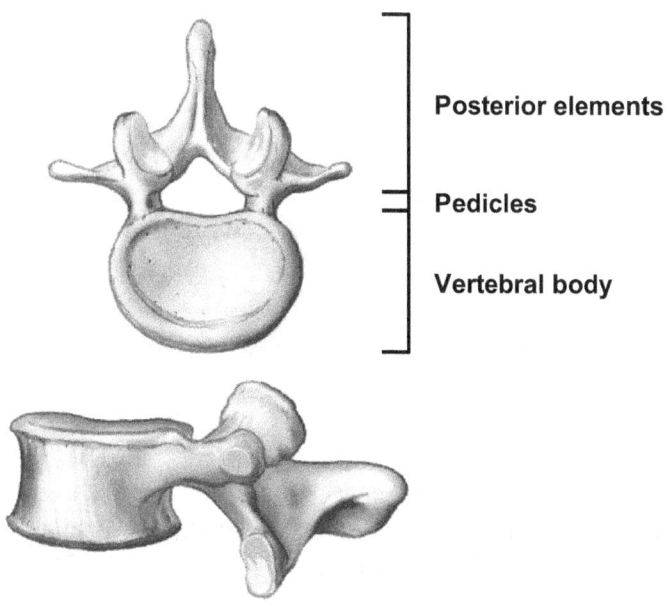

Figure 3.—Top and side views of lumbar vertebra.

Vertebral Endplates

On the top and bottom surface of every vertebral body, between the bone and the disc, there is a layer of cartilage about .04 in thick called the vertebral endplate (Figure 4). This endplate serves as an interface between the vertebral bone and the disc and plays an important role in disc nutrition. Discs do not have a direct blood supply, which is the normal method of supplying nutrition to body tissues. Thus, the health of the disc is dependent on the flow of nutrients from the interior of the vertebral body (which does have a blood supply) through this endplate. The integrity of the vertebral endplate seems to play an important role in the development of disc degeneration and LBP.

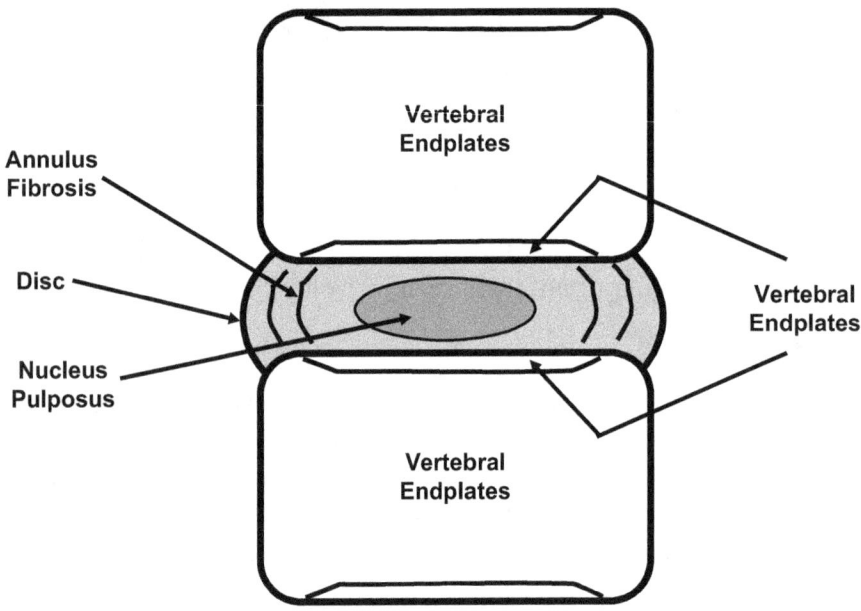

Figure 4.—Vertebral endplates are located on the vertebral body adjacent to the discs.

Intervertebral Discs

Intervertebral discs are found between each vertebra (Figure 4). The discs are flat, round structures with tough outer rings of tissue called the annulus fibrosis. These contain a soft, jellylike center called the nucleus pulposus. As mentioned previously, the disc allows the spine to be more flexible, but it also works as a shock absorber. However, when discs start to degenerate, they can become a major source of LBP. The intervertebral disc is fixed in size, and nearly all of the weight-bearing forces are distributed across its area. Therefore, the strength of the disc is limited by the applied force per unit area.

Muscles and Tendons

Numerous muscles surround the spine and are responsible for permitting it to adopt an upright posture and providing for its motion (Figure 5). The back muscles generate very large forces when lifting objects and these forces can create great stress on the lumbar spine. Tendons attach muscle to bone. When a strained ("pulled") muscle is experienced, usually the failure occurs where the tendon and the muscle join.

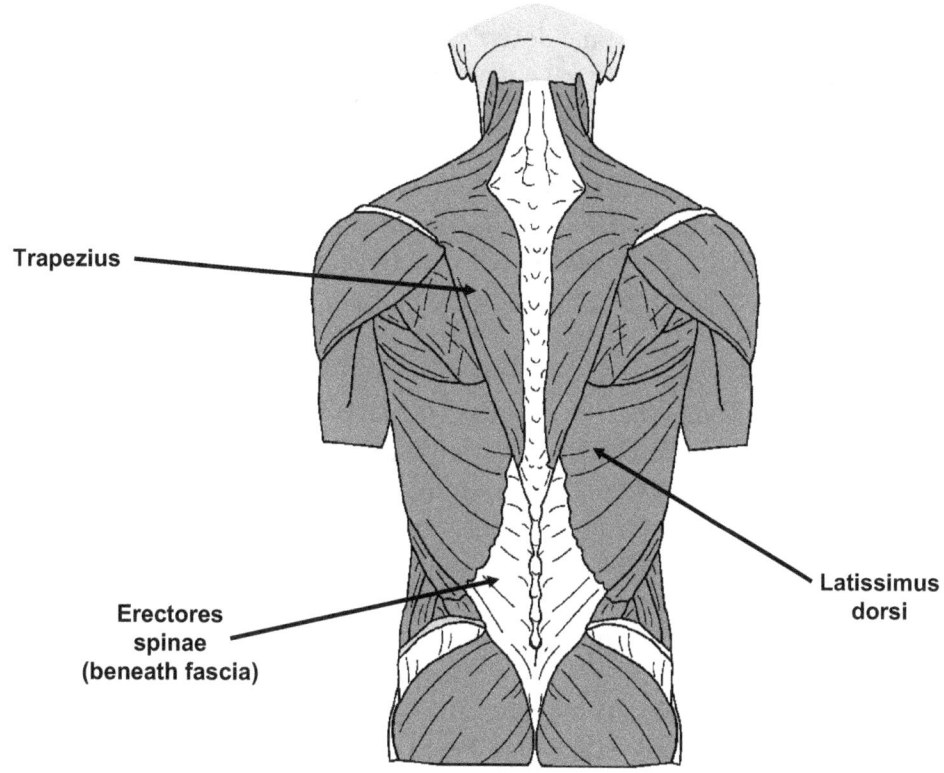

Figure 5.—Muscles of the low back.

Ligaments

Ligaments connect bone to bone. Like tendons, these structures are connective tissues made of densely packed collagen fibers. In the spine, many ligaments are available to provide structural stability. Figure 6 shows some of the major ligaments of the spine.

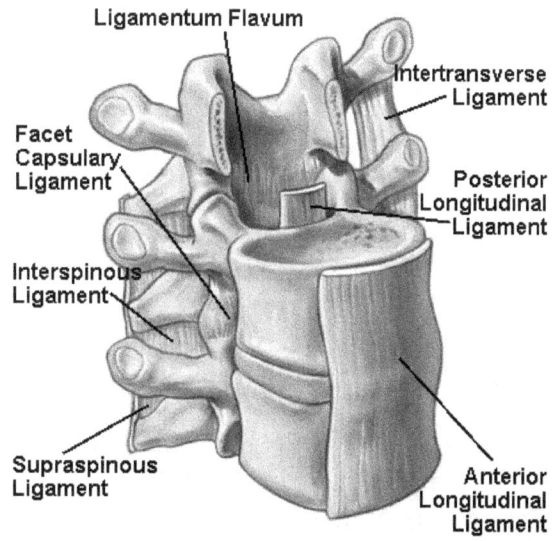

Figure 6.—Ligaments of the lumbar spine. Ligaments attach bone to bone and help provide stability to the spine.

Sources of Low Back Pain (Pain Generators)

Almost every structure of the spine described above has been a suspected source of LBP. Popular theories as to the source of LBP have included back muscle strains, ligament sprains, and so-called trigger points [Bogduk 1997]. While some of these structures may be involved in short-term LBP or discomfort, there is not much evidence to support that these structures are the source of the chronic LBP that leads to high cost and long-term disability cases. Evidence now suggests that a great deal of chronic LBP is due to disc degeneration, facet joint damage, and/ or dysfunction of the sacroiliac joint. The following sections detail how these structures and processes are believed to be related to the development of chronic LBP.

Sources of Chronic Pain

Disc degeneration. The process of disc degeneration is now thought to be a major culprit in the development of chronic LBP. Disc degeneration is believed to be initiated through endplate fractures [Adams et al. 2000]. The endplate (not the disc) is the weakest link in the spine and is the first structure to fail when subjected to a high (or repeated) load (Figure 7) [Bogduk 1997]. The endplate develops fissures or "cracks" when its strength is exceeded. The body repairs these cracks using its own particular brand of "mortar." The mortar used to repair these endplate cracks is called scar tissue. Scar tissue helps increase the structural integrity of the crack; however, it does so at a price. Specifically, it reduces the flow of nutrients from the vertebral body to the disc. When the reduced nutrition is no longer sufficient for disc health, the process of disc degeneration is set in motion [Adams and Dolan 1995].

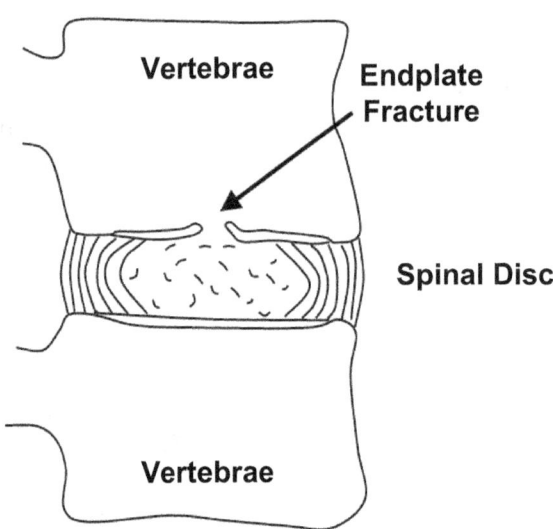

Figure 7.—The endplate is the first structure to become damaged when subjected to a high or repeated load.

Discs tend to degenerate from the inside out [Bogduk 1991]. Tears will first occur in the innermost fibers of the disc. Failure of these fibers will place additional strain on adjacent fibers, which (being similarly weakened) will also begin to fail. Once the process has been started, repeated loading will cause additional tears of the annular fibers, resulting in formation of significant disc fissures (Figure 8) [Moneta et al. 1994; Schwarzer et al. 1995b; Ito et al.

1998]. If these fissures approach or occur in the outer portion of the annulus, a wound-healing process is initiated that involves an inflammatory response. Inflammation is an important early stage of the body's repair process. However, many of the chemicals involved in the inflammation process irritate pain receptors that are present in the outer portions of the disc (pain receptors are not found in the inner portions of the disc because of the high pressures present there) [Bogduk 1997]. In patients experiencing chronic LBP, the occurrence of internal disc disruption is at least 39% [Schwarzer et al. 1995b], which represents the most common cause of chronic LBP that has yet been demonstrated [Bogduk 1997]. Herniated discs, in which the nucleus pushes outside the annulus, are fairly rare and occur mostly in slightly degenerated discs [Gallagher 2002]. Loads imposed on the spine during lifting tasks are thought to be sufficient to cause fractures in the endplates of lumbar vertebrae, particularly upon repeated loading [Brinckmann et al. 1988; Gallagher et al. 2005, 2007; Hansson et al. 1987].

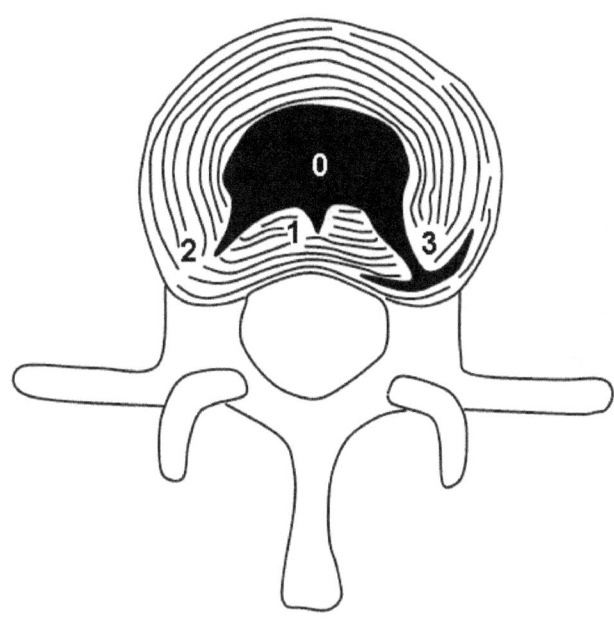

Figure 8.—Fissures (tears) in the disc. Numbers refer to grades of fissures. Grades 0, 1, and 2 are not associated with chronic pain. However, grade 3 fissures (reaching or occurring in outer portion of the disc) are highly associated with chronic LBP [Bogduk 1997].

Mechanisms of endplate fracture. The best approach to preventing the process of disc degeneration seems to be minimizing the likelihood of the initial endplate fracture. Endplate fractures can occur in two basic ways. The first is that the endplate's strength may be exceeded by one very large load placed on the spine. This may occur in the case of a fall, an extreme jolt when riding in a vehicle, or even during an extremely heavy lift. However, most experts believe that endplate fractures occur more commonly from the process of repeated loading of the spine [Brinckmann et al. 1988]. In this process, a load placed on the spine (perhaps from a heavy lift) results in a tiny fracture in the endplate or in the bone supporting the endplate. This small fracture creates an area of weakness which, with repeated lifting, will cause the microfracture to expand, ultimately leading to a large fracture. Thus, repetitive (submaximal) loading can eventually lead to an injury that is equivalent to an injury experienced in a one-time overload of the tissue strength [Brinckmann et al. 1988; Gallagher et al. 2005, 2007].

One important implication of the cumulative damage from repeated loading is that people can be performing tasks that they believe are safe (lifting a 50-lb bag), but each lift may be leading to a slight amount of damage. The resulting accumulation of weakness in the spine may lead to an injury that may result from what seems like a fairly innocuous task (like bending down to pick up a pencil), but which is really the result of damage that has accumulated over time.

Facet joints. Facet joints (see Figure 2) are a small set of joints located at the back of the motion segment. These joints are believed, when loaded, to develop wear and tear of the gliding surfaces of the joint, which may lead to arthritis over time [Eisenstein and Parry 1987]. Bending forward and/or side to side are believed to place damaging stress concentrations on the cartilage surface of the joints, leading to degeneration [Farfan 1973]. As the facet joint begins to degenerate, small defects in the joint surfaces appear. As with the disc, the cartilage surfaces of the facet joints have limited blood supply, which limits the body's ability to repair the damage. This limited repair capacity makes the facet joint susceptible to degeneration. In fact, degeneration of the facet joints is believed by some to be a result of disc degeneration [Vernon-Roberts and Pirie 1977]. This is because when a disc degenerates, it loses height. When this occurs, it causes the bone and cartilage of the facet joint (normally separated) to come into contact and grind against each other, causing degeneration of the surfaces. Studies have suggested that approximately 15% of chronic LBP cases may be due to facet joint pain [Schwarzer et al. 1994]. Since these injuries may be directly related to disc degeneration, it is possible that over half of chronic LBP cases are related (directly or indirectly) to the process of disc degeneration.

Sacroiliac pain. One final source of chronic LBP deals with mechanical disorders of the sacroiliac joint (where the sacrum of the spine joins the iliac bone in the pelvis). Evidence suggests that such disorders may be present in 15% of chronic LBP sufferers [Schwarzer et al. 1995a].

Sources of Short-term Low Back Pain

Muscle strain. There is little doubt that back muscles can be a source of pain. There are many ways in which muscles can be a pain source, but by far the most common is believed to be muscle strain [Armstrong 1990]. When muscles become overstressed in certain ways (e.g., a severe or sustained exertion), muscle fibers can become damaged usually at or near the point where the muscle and tendon join. Recent research on both laboratory animals and humans has demonstrated that muscle strain occurs most often when the muscle fibers are contracting but the muscle itself is lengthening, in what is known as an "eccentric" or "negative" contraction [Cutlip 2006]. An example of an eccentric contraction would be when a worker lowers a load to ground level. The back muscles are activated to maintain control of the load, but the muscles themselves lengthen as the back bends forward. As with other body tissues experiencing damage, repair is initiated through the pain-inducing process of inflammation.

The time required to repair the damaged tissue will of course depend on the extent of the damage. However, as opposed to other tissues that have been discussed previously (such as the intervertebral discs), the muscles have the benefit of a good blood supply. This helps significantly with the repair process, meaning that recovery from muscle strain is likely to be much more rapid than with other tissues. Interestingly, research has suggested that prolonged use of over-the-counter pain medications (such as ibuprofen), while reducing pain and inflammation, can significantly lengthen the time it takes to complete the repair process [Mishra et al. 1995]. The reason may be that while these drugs reduce inflammation and pain, the inflammation is an important part of the repair process. It has been suggested that use of anti-inflammatory drugs should be limited as a result in the treatment of muscle strains (perhaps using them only for a day or two after the initial injury).

Ligaments. When a full stooping posture is adopted (Figure 9), back muscles actually become totally inactive (i.e., stop contracting), and the ligaments of the spine have to bear the entire load of the upper body [Floyd and Silver 1955; Basmajian and DeLuca 1985]. Ligaments were once thought to be relatively simple structures and basically inert. However, in the past couple of decades it has been discovered that ligaments contain a feedback mechanism to nearby muscles. As a result, if ligaments are stretched or damaged, the strength and function of muscles can be affected. Studies have shown that spine flexion (stooping) for a 30-min period may be associated with decreased strength and back muscle spasms for a period of up to 24 hr [Solomonow et al. 1998]. Studies have further shown that spinal ligaments can and do get damaged during flexion or when the muscle-tendon unit fails. Ligaments are slow to heal because of a poor blood supply [Solomonow 2004]. This may decrease their ability to properly support the spine and may lead to decreased muscular function as well. This is one of many reasons that the stooping posture is hazardous to spinal structures and should be avoided as much as possible. The role of ligaments in LBP is very relevant to the underground mining environment, especially in low- to midseam mines where workers frequently adopt a stooping posture.

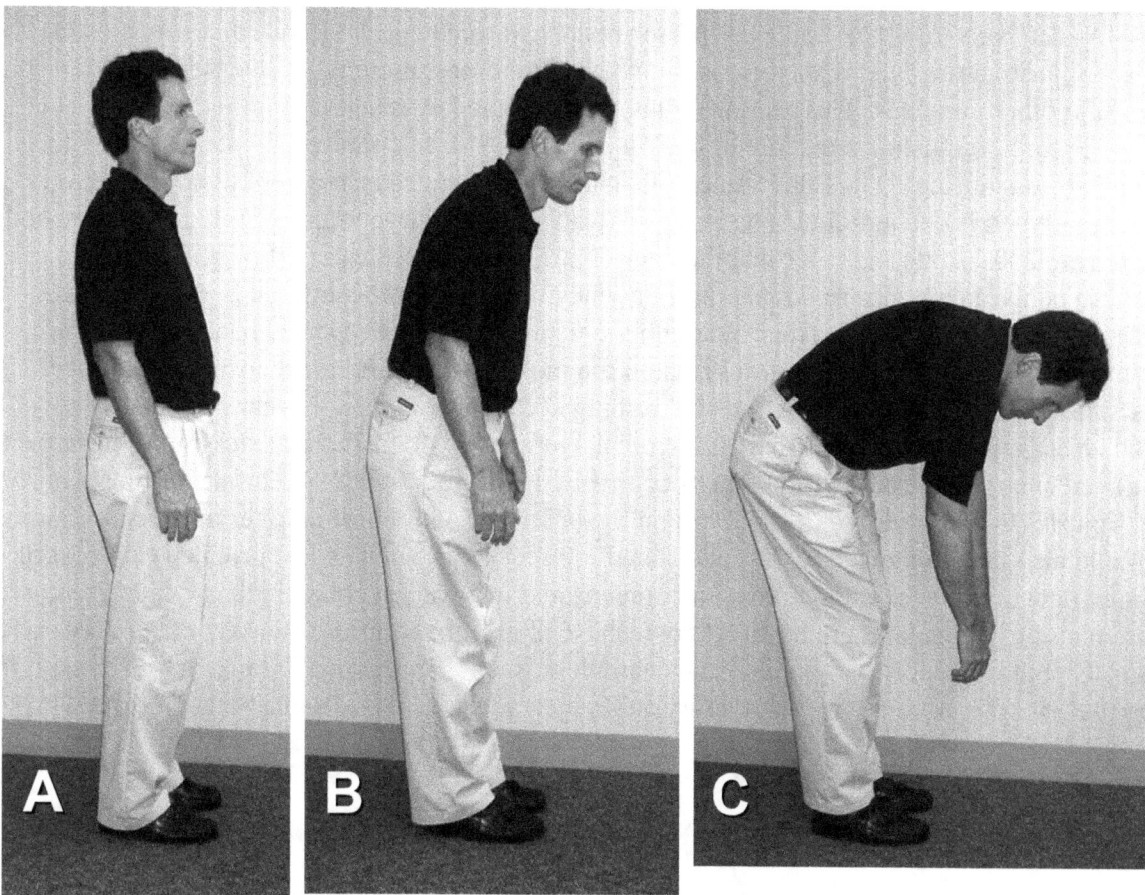

Figure 9.—Adopting a full stooping posture. Back muscles: (A) exhibit a small amount of activity in standing, (B) exhibit increased activity in slight bending, and (C) become inactive when stooping. Ligaments bear the weight of the upper body in the stooping posture.

II. LOW BACK DISABILITY IN MINING

It was not all that long ago that coal and ore from underground mines were shoveled or manually loaded onto carts drawn by horse or mule [Sanders and Peay 1988]. As recently as the mid-1950s, almost one-third of all coal produced in the United States was still hand-loaded. In years prior to mechanization, mining was truly back-breaking work. Advances in mechanization in the second half of the 20th century have greatly reduced physical demands on the mine worker. Even so, mining remains a very physically demanding occupation, and many unique environmental difficulties remain. For example, miners may have to deal with severely restricted workspace [Lineberry and Adler 1988], poor illumination, muddy or wet ground or floor conditions, high levels of WBV, and considerable heavy lifting. The significant LBP and disability experience in the mining industry is undoubtedly the consequence of the numerous risk factors present in the mining environment. This section will examine recent trends with respect to low back disability in mining and will discuss working conditions unique to mining that may contribute to the problem of LBP and disability, particularly work in awkward or restricted postures. Research related to the risks and physical limitations of restricted postures will be presented and discussed.

Reportable Back Injuries in Mining (1996–2005)

An analysis of back injuries reported in the Mine Safety and Health Administration (MSHA) database for 1996–2005 was performed to examine current trends and statistics related to back injuries associated with various mining commodities, mine types, worker activities, job titles, and other variables. This section details some of the key results from this analysis.

Figure 10 illustrates the rate of back injuries per 100 full-time equivalents (FTEs) for all mining sectors during 1996–2005. A decline in back injury rates is apparent over this period. While this trend is encouraging, a separate analysis of cumulative injuries broken down by body part during 2003–2004 indicates that the back is still the predominant part of body affected (31%) in coal mining, almost double the frequency of the next most injured body part (knees at 17%) [Moore et al. 2007]. Thus, although the incidence rate of reported back injuries has declined in recent years, back injuries remain the most frequent and most costly cumulative musculoskeletal injury, at least in the coal mining sector. During the 10-year period studied, 25,607 back injuries were identified. Given that average direct costs for a back injury have been estimated at approximately $18,000 per occurrence [Mitterer 1999; NSC 2004], the costs of back injuries to the mining industry were on the order of $460.9 million over the decade studied. It should also be noted that LBP has additional indirect costs, such as lost productivity, cost of training new or replacement workers, uncompensated lost wages, etc., which can increase costs to double or triple the amount of direct costs [NRC 2001]. In addition, many workers may suffer from LBP but not report it as an injury. Nonetheless, such "nonreported" pain may also result in decreases of both productivity and quality and may lead to increased safety-related risks.

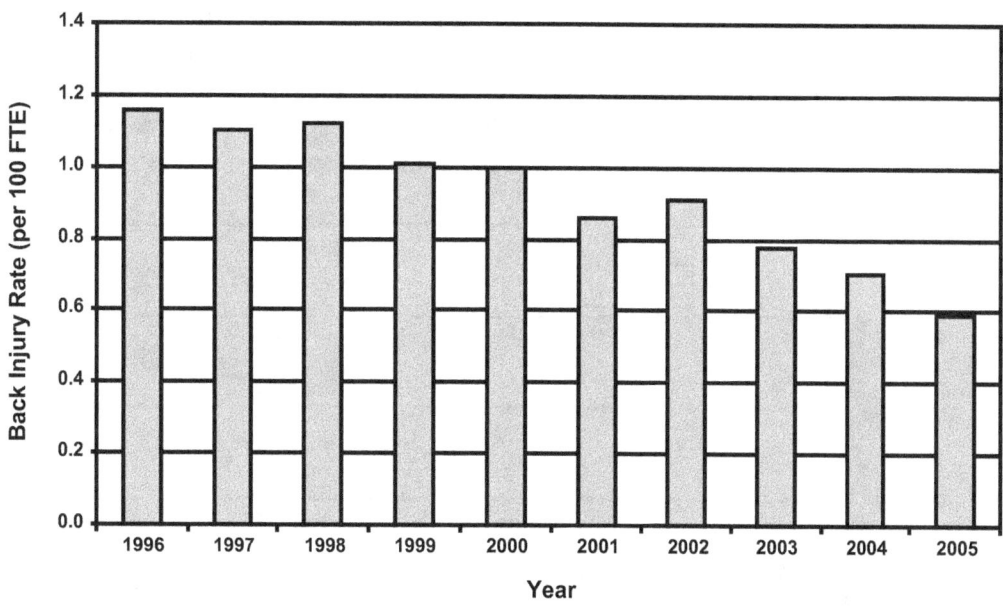

Figure 10.—Back injury rates per 100 FTEs over all mining commodities and contractors (1996–2005). (Source: MSHA database)

Examining the injury trend in underground versus surface mining (Figure 11) reveals that while both types of mining have experienced decreases in back injury rates over the 10-year period studied, underground mining has a consistently higher rate of back injuries compared to surface operations. In fact, the overall back injury incidence rate for underground mining (1.79 back injuries per 100 FTEs) was more than double that observed for surface mining (0.74 per 100 FTEs).

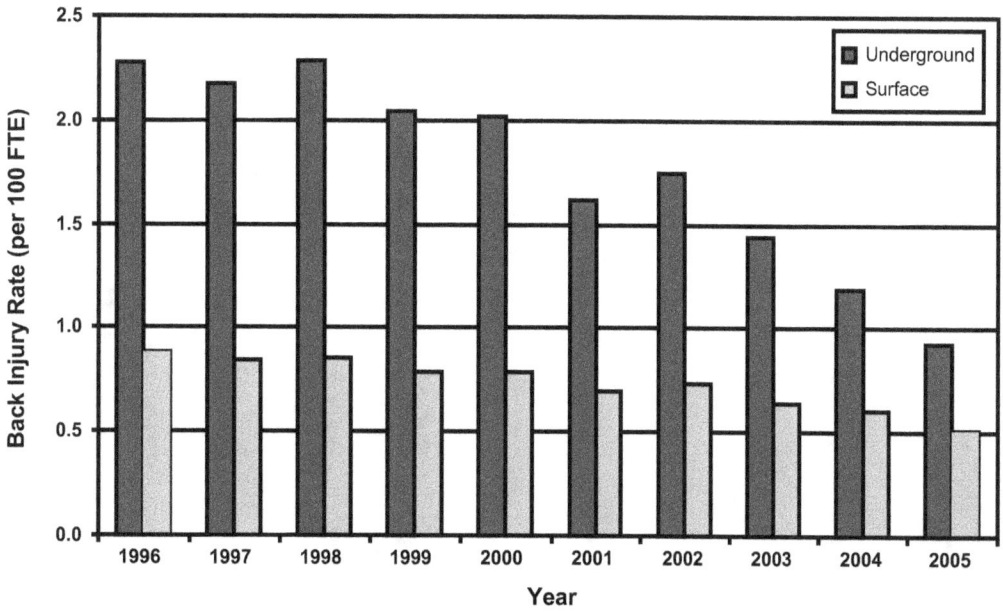

Figure 11.—Back injury rates per 100 FTEs, underground versus surface (1996–2005). (Source: MSHA database)

Figure 12 presents back injury rates by commodity and mine type (underground versus surface) during 1996–2005. It is apparent from the bar graph that the risk of back injury varies with respect to different commodities. Underground coal mining stands out as the most vulnerable sector in the mining industry in terms of the rate of workers experiencing back injuries. This is true for both operators and contractors. Figure 12 also illustrates the tendency for underground operations to experience higher rates than surface operations.

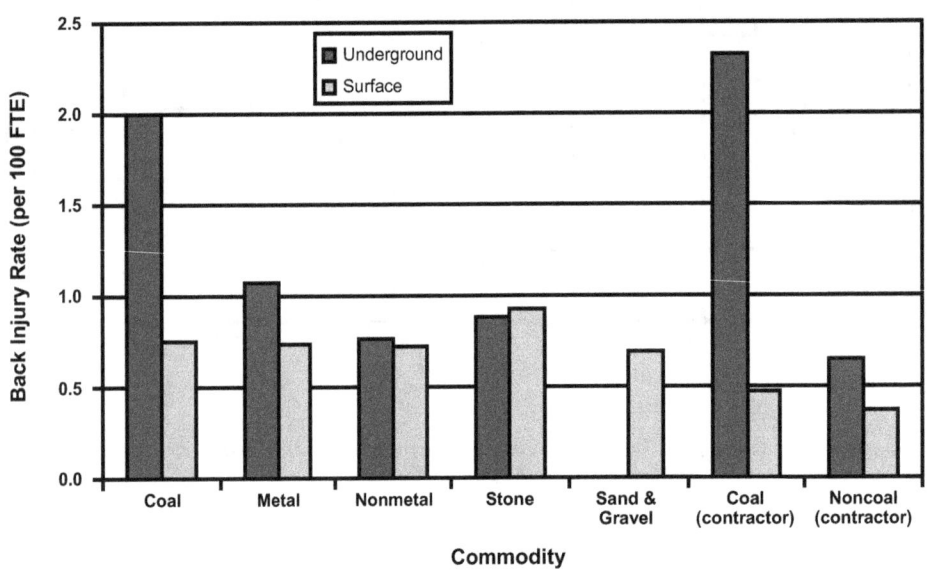

Figure 12.—Back injury rates by commodity and mine type (underground versus surface) aggregated from 1996 to 2005. (Source: MSHA database)

Figure 13 shows the percentage of back injuries by mine worker activity performed at the time the back injury occurred (1996–2005). By far, the largest proportion of injuries was classified as handling materials (31.3%). However, other categories, such as moving power cable and hand loading/shoveling (each responsible for approximately 4.5% of the total), also involve material handling, and their inclusion would increase the total percentage of back injuries resulting from material-handling activities to at least 40%. Other common activities associated with back injuries include walking and getting on and off equipment (where trip, slip, and fall accidents would be expected to be common), machine maintenance, and use of handtools.

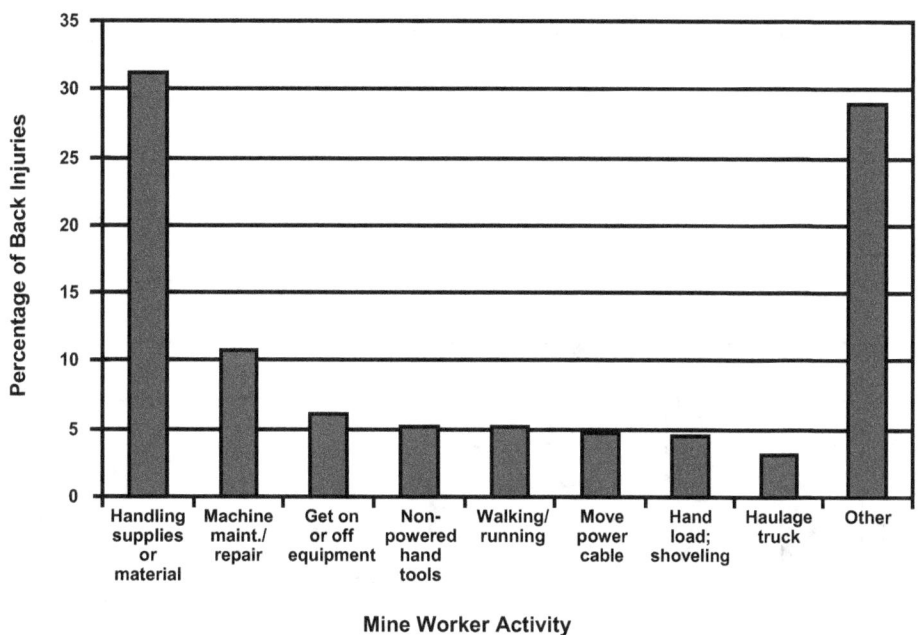

Figure 13.—Percentage of back injuries by mine worker activity (1996–2005). (Source: MSHA database)

Figure 14 shows a breakdown of back injuries in mining during 1996–2005 by the seven leading job titles in terms of injury frequency. In this analysis, similar job titles (such as different categories of laborer) were grouped together. One can see from this bar graph that the laborer job title represents about 15% of back injuries, followed by mechanic/repairman (13%), cleaning plant operators ("boney" refers to waste removed in a cleaning plant) (10%), truck drivers (5.5%), heavy equipment operators (4.3%), roof bolter operators (4.2%), and high lift operators/front-end loaders (4.0%). By and large, these job titles represent workers who either perform heavy manual work, probably involving materials handling and physical exertion that puts a high load on the low back, or are exposed to a significant amount of WBV when operating trucks and heavy industrial equipment (which is also related to the development of back injuries).

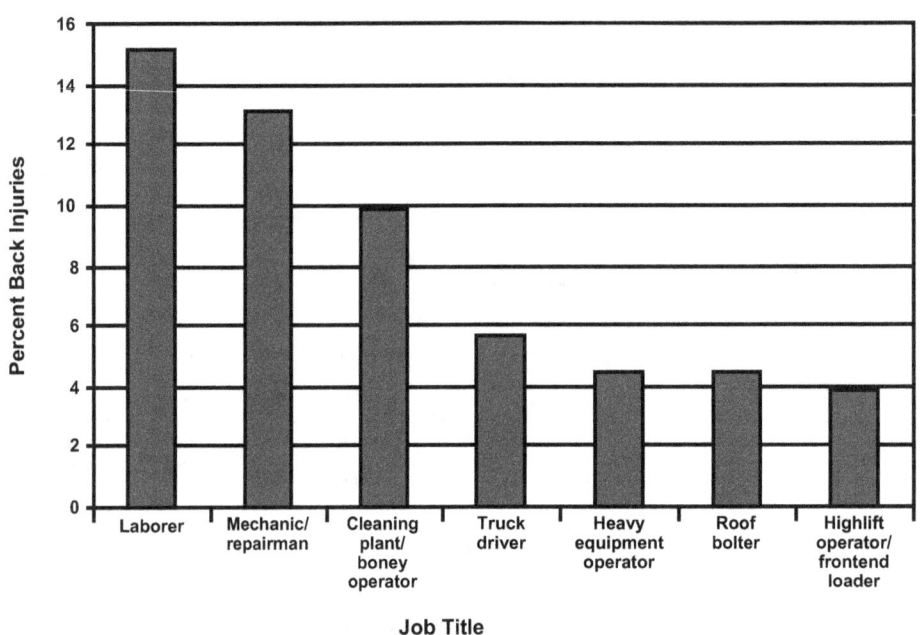

Figure 14.—The seven leading job titles in terms of the percentage of back injuries reported during 1996–2005. (Source: MSHA database)

This analysis discloses several results of great relevance to topics covered in this report. One remarkable finding is the decline in incidence rates of back injury observed across all sectors of the U.S. mining industry. This may be a result of increased attention being paid to back injuries by the industry in recognition of the tremendous cost of these injuries. Mining companies are beginning to institute ergonomics programs to reduce the risk of back injuries and other musculoskeletal disorders, which may be having a positive impact on these rates. Many companies have hired physical therapists and rehabilitation professionals to assist with control of these injuries. More attention is being paid to providing well-designed seats to mine workers throughout the industry, which may mitigate risk of back injury from exposure to WBV. Nonetheless, back injuries remain the most significant source of disability, cost, and suffering of all musculoskeletal disorders in mining, and it is clear that there is much work to do in this area [Moore et al. 2007].

Unique Demands of the Mining Environment

One aspect of the underground mining environment that increases LBP risk is the severe restriction in workspace present in many underground mines. Restricted space forces mine workers to adopt awkward work postures that both affect their physical work capacities and the loading experienced by the low back [Gallagher 2005]. While the human body is still capable of performing work tasks in such environments, it cannot perform equally as well as in unrestricted conditions. In fact, when faced with awkward tasks or environmental demands (like those present in the mining environment), workers may endure substantial performance limitations. However, workers may not always be aware of the risks or limitations imposed by these postures.

Numerous mining jobs require workers to operate in less desirable postures, such as kneeling, stooping, squatting, and/or lying down [Haselgrave et al. 1997]. Such postures are common for mechanics or maintenance personnel at surface mines and for any underground worker in low- to midseam coal mines. In the past couple of decades, researchers have examined the adaptations, limitations, and tradeoffs associated with working in these restricted or awkward postures. This research is important in assessing risk of back injuries in these postures and appropriate task design for restricted postures.

Low Back Pain and Restricted Postures

Restricted postures (especially stooping and kneeling) seem to be associated with an increased risk of LBP. However, it should be noted that adoption of awkward postures often coincides with other risk factors, such as lifting. The overlap in risk factors makes it problematic to determine exactly what is the risk associated with lifting versus the risk of the posture used during the lift. It may be the combination of lifting in awkward postures that poses the most significant risk for workers.

Lawrence [1955] examined British coal miners to identify factors related to degenerative disc changes. Lawrence found that injury, duration of heavy lifting, duration of stooping, and exposure to wet mine conditions were the factors most associated with spinal changes. The finding of increased low back compensation claims in conditions where stooping predominates is congruent with other evidence relating nonneutral trunk postures to low back disorders. For example, a study by Punnett et al. [1991] examined the relationship between flexed trunk postures and risk of low back disorders. After adjusting for factors such as age, sex, length of employment, and medical history, time spent in flexed trunk postures was strongly associated with back disorders.

A study of 1,773 randomly selected construction workers also examined the effects of awkward working postures on the prevalence rates of LBP [Holmström et al. 1992]. This study found that prevalence rate ratios for LBP were increased for both stooping and kneeling when the duration of work in these postures was reported to be at least 1 hr per day. Furthermore, a dose-response relationship was observed whereby longer durations of stooping and kneeling were associated with increased prevalence rate ratios for severe LBP. Thus, workers who adopt stooping or kneeling postures for longer periods of time seem to be at increased risk of experiencing severe LBP. The increased risk of LBP in restricted postures may be the result of reduced physical capabilities and increased loads on the spine, as detailed in the following sections.

Work in Restricted Postures

Workers enjoy the benefits of high-strength capabilities and superior mobility when they adopt a normal (neutral) standing position. Standing maximizes the ability to use major muscle groups of the body in contributing to the forces needed to accomplish a task. This posture also permits all body segments to participate in a task and is a position where the length of muscles is at or close to optimal in terms of force production. However, this muscular synergy can be badly disrupted when unusual or restricted postures are used. Some restricted postures reduce the number of muscles that can be called upon to perform a task. In others, muscles may be in a poor position in which to generate the needed forces. One need only imagine performing a lift while lying down on the ground to understand that many powerful muscles (i.e., those of the legs, hips, and thighs) are unable to effectively participate in the lifting task. The result is that other muscles may incur additional burdens in performance of the task and may become overstrained as a result. Each unique postural configuration will result in its own set of strength limits. The number and identity of the muscles that can be effectively recruited for the job will largely determine these limits [Dul 1986].

Task performance in restricted work postures can also be affected by reduced mobility, stability, and balance. For example, if a worker is unable to stand on his or her feet, mobility will be dramatically reduced. This factor can have a significant impact on methods used for task performance. Consider lifting an object from directly in front of the body to a position to the side of the body as performed in two postures: standing and kneeling. When a worker is standing, it is reasonable to request that the worker avoid twisting the trunk simply by repositioning the feet. However, the task of repositioning is considerably more difficult when kneeling (especially when handling a heavy load), and workers are not inclined to take the time nor the effort to do this. Instead, the worker opts for the quicker trunk-twisting method, which can place a high stress on the low back. Maintaining balance may also impact task performance in restricted postures. Workers may have to limit force application in certain postures in order to maintain balance (particularly in kneeling and squatting postures).

As mentioned previously, awkward work postures are often the consequence of restrictions in workspace, typically in vertical or lateral dimensions, or poor workspace design. Workspace restrictions of this sort put not only the worker in a bind, but anyone seeking to intervene to reduce job demands (an ergonomist, for example). The worker is restricted by the limitations of the posture he or she must employ. Those trying to redesign the job may be deprived of favored techniques for reducing musculoskeletal disorder risk. For example, restricted space greatly limits the number and type of mechanical devices (cranes, hoists, forklifts, etc.) available to reduce the muscular demands on the worker. If mechanical assistance is to be provided, it frequently must be custom-fabricated for the environment. Restrictions in workspace also limit opportunities to ease the strain arising from the worker's postural demands, often forcing the ergonomist to recommend working postures from a limited menu of unpalatable alternatives.

Restricted spaces can also result in more subtle effects. One is the tendency, as vertical space is reduced, to force workers into asymmetric (twisting) motions. Lifting straight in front of the body is generally preferred in the standing posture, but becomes progressively more difficult if one is stooping in reduced vertical space. In fact, lifting capacity in asymmetric lifts (i.e., lifts to the side) tends to be higher than when lifting directly front of the body in low-seam heights [Gallagher 1991]. This represents a change from the unrestricted standing position, where asymmetry reduces lifting capacity [Garg and Badger 1986]. Finally, as Drury

[1985] points out, space limitations tend to impose a single performance method on a worker. In unrestricted spaces, when a worker's preferred muscles fatigue, it is often possible for an individual to employ substitute motions that may shift part of the load from fatigued muscles. In a 39-in coal seam, the worker will have to perform lifting tasks in a kneeling posture, without much opportunity to relieve the muscles that need to be used in this posture. The likely result is intensified fatigue, decreased performance, and increased risk of injury in restricted postures.

Restricted Postures and Lifting Capacity

Studies have made it clear that the lifting capacity of individuals is reduced when restricted postures are adopted compared to standing. The amount of the decrease varies according to the posture used. In general, studies have found that a kneeling posture is associated with a 7%–21% decrease in lifting capacity compared to the standing posture and does not seem to differ much if lifting on one or both knees [Ayoub et al. 1985a,b; Gallagher and Unger 1990]. The sitting posture results in a similar, but slightly greater decline, averaging 16%–23% less than in the standing posture. A more substantial decline in lifting capacity is seen in squatting, ranging 20%–33% less than standing. Lifting in a lying-down posture is associated with drastic reductions in lifting capacity—up to 75% less than what could be lifted in a standing position [Ayoub et al. 1985a,b].

One-lift maximum acceptable lifting capacities for male and female subjects are presented in Figures 15 and 16, respectively. The data shown in these figures were obtained from U.S. Air Force maintenance personnel who performed lifting tasks in standing, kneeling (on one and both knees), sitting, and squatting postures [Ayoub et al. 1985a,b; Gibbons 1989]. It is apparent from the data that there is a reduction in lifting capacity in kneeling, sitting, and squatting postures and that the effects of posture on lifting capacity are more pronounced with lifts of 35% of vertical reach (i.e., the highest point that could be reached by an individual in a given posture). The effect becomes progressively diminished (though still apparent) when lifts to 60% and 85% of vertical reach are performed. It may be that strength capabilities for lifts to higher heights may be controlled more by limitations in shoulder and arm strength and are thus not as dependent on body posture. Finally, comparison of male strength (Figure 15) versus female strength (Figure 16) indicates that the effects of posture are similar for both sexes. However, the strength exhibited by females averaged about 50%–60% of that achieved by their male counterparts (note that the Y-axis scale is different for Figures 15 and 16).

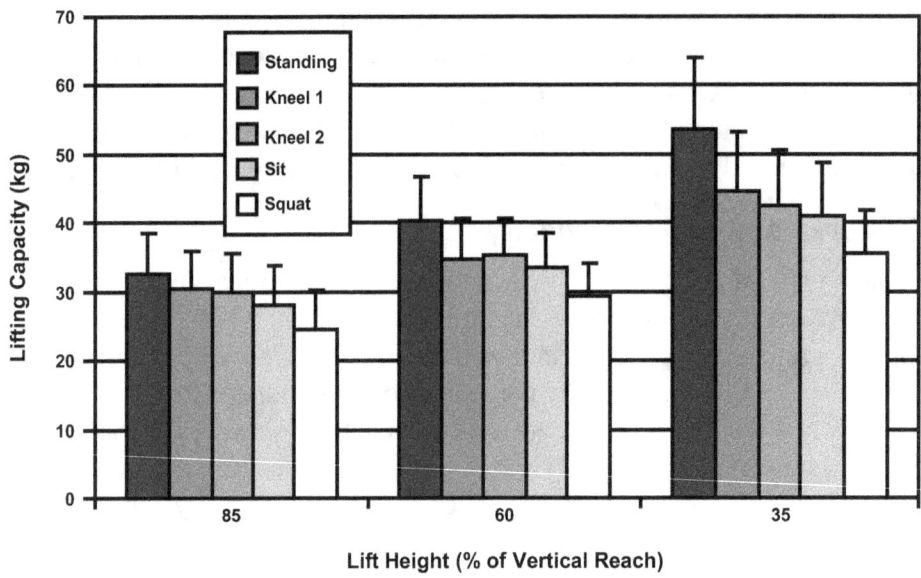

Figure 15.—One-lift maximum lifting capacity for males in various postures [Ayoub et al. 1985a].

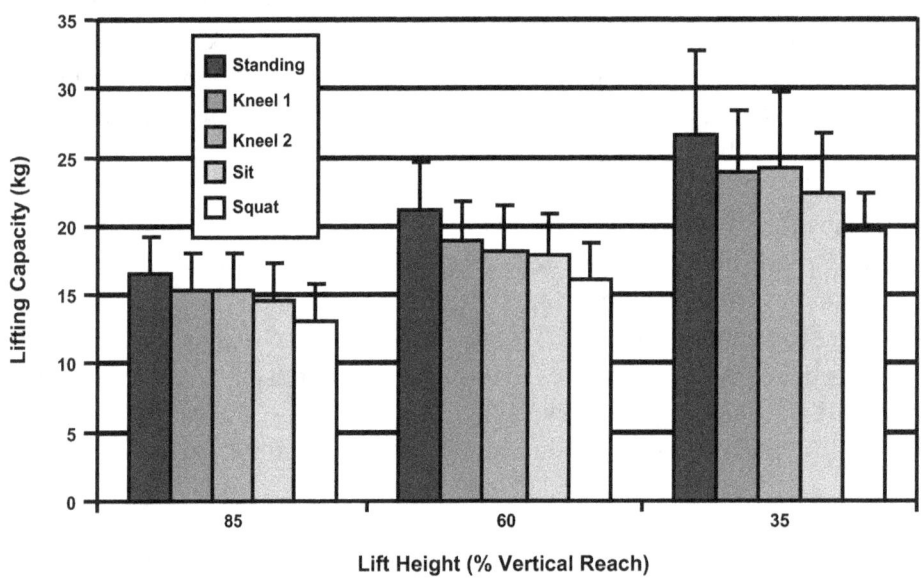

Figure 16.—One-lift maximum lifting capacity for females in various postures [Ayoub et al. 1985a].

It should be emphasized that the data discussed in the previous section represent the amount of weight that was acceptable to lift one time and assumes that workers would perform such tasks only occasionally, not for extended periods. However, periods of more extended lifting in restricted posture are common in mining. Thus, several studies examining the lifting capacity of underground coal miners adopting restricted postures involving repetitive lifting activities have been conducted [Gallagher et al. 1988; Gallagher and Unger 1990; Gallagher 1991; Gallagher and Hamrick 1992]. These studies also used the psychophysical approach, allowing subjects to adjust the weight in lifting boxes to acceptable loads during 20-min lifting periods. These studies examined lifting capacities in standing, kneeling, and stooping postures.

Results of these studies show lifting limitations similar to those of the studies described previously. Restricted postures (stooping and kneeling) were found to result in 5%–25% lower estimates of acceptable loads compared to the standing posture [Gallagher and Hamrick 1992]. Kneeling was found to have a significantly reduced estimate of acceptable load (13%–20%) compared to stooping [Gallagher et al. 1988; Gallagher and Unger 1990; Gallagher 1991]. While posture was almost always an important determinant of lifting capacity in these studies, there were some factors that, if present, could reduce or eliminate the effect. In particular, it was found that if items had a poor hand-object coupling (no handholds), lifting capacity could be reduced to such an extent that effects due to posture were no longer an issue [Gallagher and Hamrick 1992].

Effects of Restricted Postures on Muscular Strength

Studies examining strength capabilities in unusual or restricted postures are relatively rare. Ayoub et al. [1981] presented an intriguing set of strength data comparing isometric strengths of coal miners working in restricted postures to a population of industrial workers (Figure 17). Strength measures included back strength, shoulder strength, arm strength, sitting leg strength, and standing leg strength. When compared with a sample of industrial workers [Ayoub et al. 1978], low-seam coal miners were found to have significantly lower back strength, but much higher leg strength. Ayoub et al. ascribed the decrease in back strength to unspecified factors related to the postures imposed by the low-seam environment. Low-seam coal miners may be obliged to work in a stooping posture for extended periods. In this posture, the spine is largely supported by ligaments and other passive tissues, "sparing" the use of the back muscles. Studies of lifting in the stooping posture suggest that the gluteal muscles and hamstrings provide a large share of the forces in this position [Gallagher et al. 1988]. The findings of Ayoub et al. [1981] may reflect a relative deconditioning of back muscles when stooping and an increased reliance on the leg and hip musculature to perform underground work tasks (producing an increase in leg strength). Further research is needed to ascertain long-term effects on strength resulting from prolonged work in restricted postures. Gallagher [1997] investigated back strength and muscle activity in standing and kneeling postures. Findings of this study showed that back strength is reduced by 16% in the kneeling posture compared to standing, similar to decreases observed in psychophysical lifting capacity when kneeling.

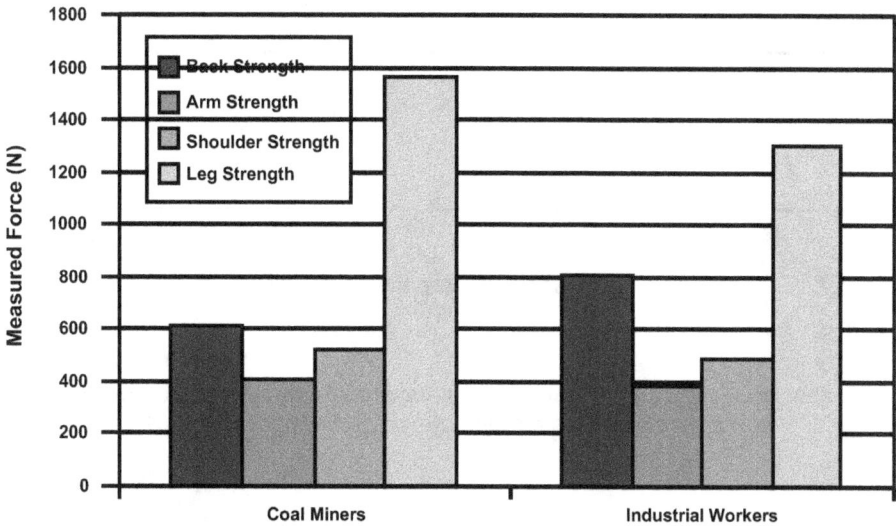

Figure 17.—Comparison of strength measures for coal miners working in confined vertical space [Ayoub et at. 1981] to an industrial population [Ayoub et al. 1978].

Effects of Restricted Space on Spine Loading

Studies have suggested that one of the best predictors for LBP is the external moment about the lumbar spine that results from the product of the force required to lift an object multiplied by the distance these forces act away from the spine [Marras et al. 1993]. As illustrated in Figure 18, recent evidence has shown that as vertical workspace becomes more restricted, the moment experienced by the lumbar spine is increased [Gallagher et al. 2001]. Of course, such a response would be expected in the standing posture, where reduced ceiling heights would cause the trunk to bend forward, increasing the moment experienced by the lumbar spine. However, this study, which involved lifting heavy mining electrical cables, found no difference between stooping and kneeling postures in terms of the peak spinal moment experienced by the subject. The primary determinant of the lumbar moment was the ceiling height. Lower ceilings were associated with higher lumbar moments and vice versa (no matter which posture was used). The question raised by this study is: why was there not a decreased moment when the kneeling posture was employed? Clearly, the trunk can maintain a more erect posture when kneeling. However, analysis of this position indicates that kneeling (with the knees in full flexion) can cause the worker to have to reach forward a considerable amount to reach a load located in front of the knees at the beginning (and most stressful part) of the lift. This creates a large horizontal distance between the spine and the load, resulting in a large moment, apparently offsetting the benefits of maintaining a more erect trunk position.

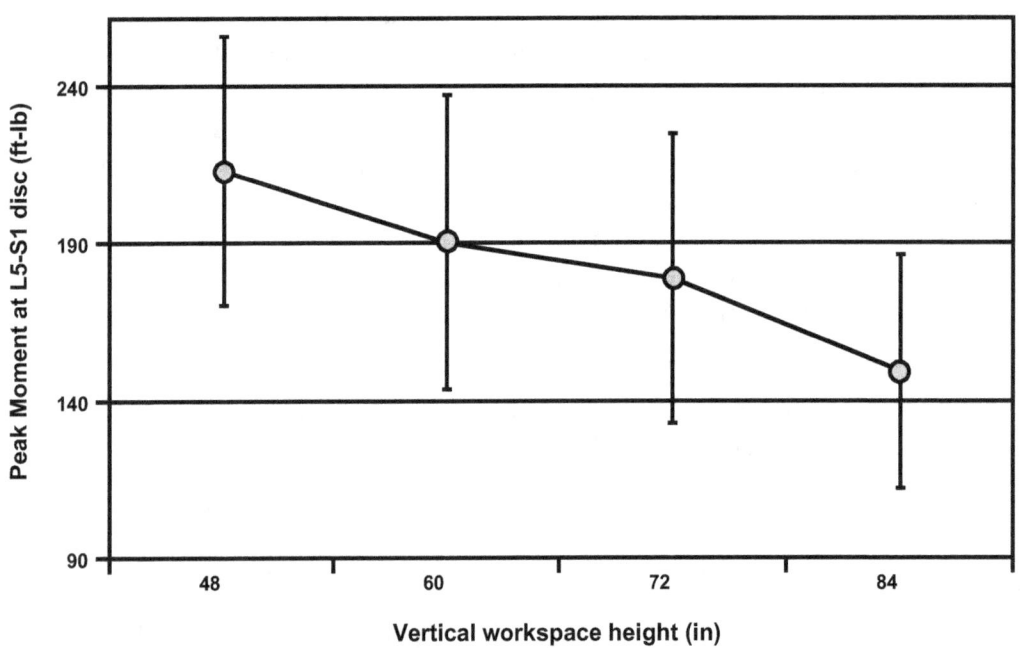

Figure 18.—Lumbar moments, an indicator of load imposed on the low back, are increased as vertical workspace becomes more confined [Gallagher et al. 2001].

Summary

In summary, research has shown that work in restricted postures is associated with increased incidence of LBP and substantial declines in the physical capabilities of workers. Lifting capacity and strength are both significantly reduced in restricted postures compared to what can be achieved when standing. Since LBP risk is increased when lifting capacity is at or near maximum capacity, this suggests that loads that must be manually handled should be reduced compared to loads lifted in an unrestricted standing posture. In general, the data suggest that a 20%–30% reduction in load should be put into effect when workers adopt restricted postures in order to reduce LBP risk due to overexertion. The finding that spinal moments continue to increase as vertical space gets more restricted also argues for lightening loads for workers in these conditions [Gallagher 1999; Gallagher et al. 2002]. Moments about the lumbar spine are one of the best predictors of the risk of LBP in performing manual handling tasks.

III. PRIMARY PREVENTION OF LOW BACK PAIN IN MINING

Primary prevention efforts for LBP (i.e., prevention of an initial episode) in mining should focus on controlling the risks associated with manual materials handling and WBV exposure. With respect to the former, elimination of manual lifting through better design of systems and use of mechanical aids can be quite effective, as can improved design of lifting tasks when manual lifting must be performed. Reducing WBV exposure requires improved design of seating and suspension systems in mobile equipment. These issues will be addressed later in this report. However, truly effective prevention efforts require an integrated and coordinated approach through which problem jobs or tasks are identified and control measures put in place prior to the development of an LBP case. In other words, a *proactive* approach is necessary [Cohen et al. 1997].

Establishing a Proactive Process for Preventing Low Back Pain

The mining industry is diverse, and methods that are effective in risk reduction at one mine site may not be readily applicable at another because of differences in mine characteristics and/or available resources. However, the steps taken to implement effective proactive strategies for controlling LBP should be similar regardless of characteristics or resources. Specifically, four critical steps of what is known as the *risk management cycle* should be followed:

(1) Hazard identification

(2) Risk assessment

(3) Control development and implementation

(4) Review and evaluation

As illustrated in Figure 19, this truly is a cycle directed toward constant improvement.

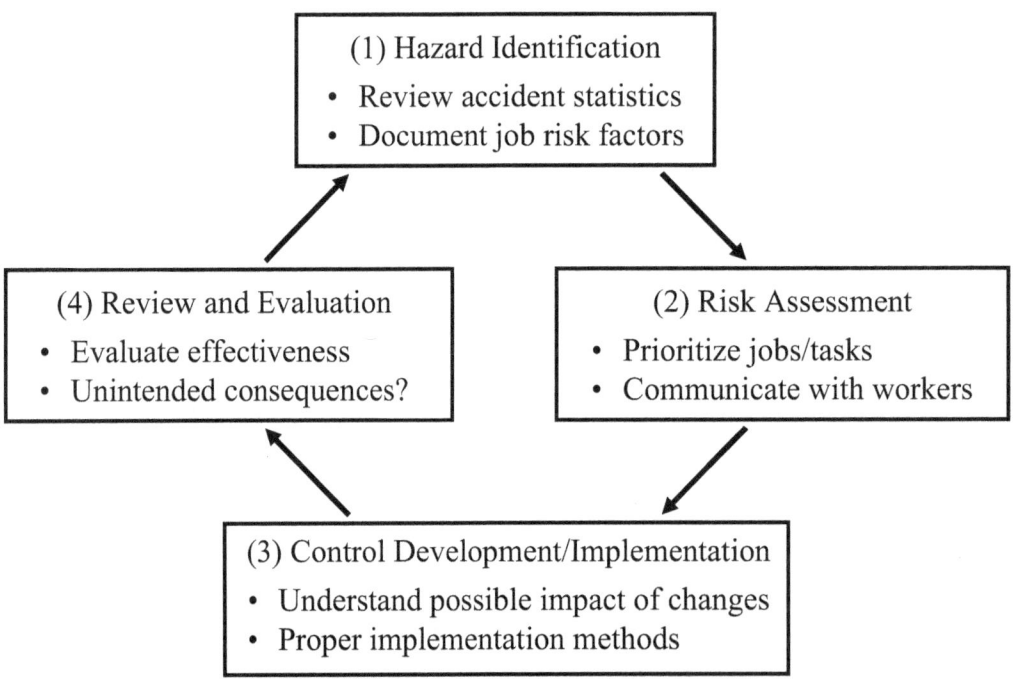

Figure 19.—The risk management cycle.

Research has shown that there are two critical pieces that must be in place to effectively implement a successful proactive approach. The first is management commitment to the process [Cohen et al. 1997]. Management must provide necessary resources in terms of time, training, policy goal development, and budgets necessary to conduct proper risk management activities. The second critical aspect is active worker participation in the process. A wide variety of individuals and a breadth of skills may be needed to be effective in addressing the various phases of the risk management cycle. A growing body of evidence indicates that worker participation is critical to success at all steps in the risk management cycle [Burgess-Limerick et al. 2007; Cole et al. 2005; Rivilis et al. 2006; Torma-Krajewski et al. 2007].

Hazard Identification

There is usually existing information that can be used to help identify jobs that pose the highest risk for LBP at a mine site. Important aspects of a systematic analysis of job risks include reviewing past accident statistics, observing the problem jobs or areas in the workplace, interviewing the workers and supervisors involved, and/or administering specific job risk factor checklists for LBP risk. Often, a great deal can be learned by examining existing documents pertaining to the job or task, and this approach can save much time in the information-gathering process. Small group discussions (called focus groups) that include the affected employees can be very helpful in gaining insight into the problems that exist with a particular job, task, or piece of equipment [Torma-Krajewski et al. 2006]. It is generally useful to engage a combination of techniques to evaluate a job and not rely on any one analysis technique. In many cases, one's own knowledge and expertise regarding the processes and procedures at the mine site may provide the best resource for identifying the problems and risks that are present.

Risk Assessment

Once the problem jobs have been identified through the evaluation processes described above, the next step is to develop a method by which to prioritize those jobs where hazards were identified. A number of factors can be considered in this process depending on the nature of the problems and the resources available to improve the jobs. A few questions that may be helpful in setting priorities include:

1. What is the seriousness (or risk) associated with the problem?

2. How many workers are affected by the identified problem, and who would benefit from providing a solution?

3. What is the complexity of the solution to the problem, what resources are needed, and how much will it cost?

An important part of the prioritization process is communication. It is important, for example, that workers realize that a visit to observe a particular job does not necessarily indicate that immediate improvements will be made. Additionally, employees who request job modifications based on what they perceive to be high-risk exposures often expect a rapid response to address their concerns. While their problems may well need to be addressed, there may be other problem jobs that are more serious or that affect a greater number of workers. Employees are usually understanding if such decisions are explained to them. However, if no such information is forthcoming, it can often lead to employee dissatisfaction with the program and expression of their displeasure to others. This can have a detrimental effect on a program that relies heavily on worker involvement to achieve success.

Control Development and Implementation

All aspects of the task or job being changed should be fully understood before making decisions regarding potential solutions. This includes understanding the effects of the proposed change on the work system as a whole and, particularly, tasks that directly precede or follow the particular task being targeted. It is often useful to develop a suite of alternative options for redesign. These may include changes in the design of equipment, tools, or supplies; work practices or administrative procedures; or workplace layout. Discussion and testing of the various proposed interventions can help determine the appropriate approach to use. Involvement of affected employees in the planning stage is crucial to acceptance of the intervention.

In the process of choosing the appropriate intervention approach, it should be noted that there is a hierarchy of controls that should be considered. The most effective and highest-priority technique is *engineering controls*. Engineering controls involve taking steps to change the job so that worker exposure to risk factors is reduced. An example would be to provide a worker with a mechanical-assist device to lift an object rather than the worker lifting it manually. *Administrative controls* may include changes to work practices or the organization of work. Examples include job rotation (rotating employees through different jobs), adjusting schedules or work pace, and/or providing recovery time for workers. The third and least effective method of reducing risk is providing personal protective equipment (PPE). Unfortunately, there is little PPE that can help prevent LBP. Back belts, for example, have been shown to have no benefit in reducing LBP [Wassell et al. 2000]. There is a steep drop in effectiveness with each successive level of the hierarchy. Engineering controls are clearly the most favored approach, and administrative controls are next best, with PPE being least effective at preventing LBP.

Once an appropriate control has been chosen, it must be effectively implemented. A thoughtful approach to implementation can greatly increase the chances of a smooth transition from the old task procedures to the new. Involving the affected employees and having them assist with the transition is an important aspect of this phase.

Interventions can be implemented in many ways. Some common methods include the following [Dul and Weerdmeester 1993]:

- *Implementation from scratch:* No old procedure exists. Training and recruitment of personnel may be necessary.
- *Direct transition:* The old procedure is completely replaced by a new one at a designated time. There is sometimes a transitional decrease in productivity immediately following implementation of this method.
- *Phased introduction:* Successive phases of the new procedures are introduced gradually. User acceptance is often more difficult with this approach due to the fact that the advantages of the system may not be apparent during the early phases of implementation.
- *Parallel application:* In this technique, both the old and new procedures are used side by side during the introduction period. Users may need to be encouraged to use the new methods due to their familiarity with the old. Apprising these workers of the advantages of the new procedures may help with acceptance.

The choice of installation strategy should be made well in advance so that planning can be done to smooth the transition. Training of all involved with the new procedures may be necessary. However, different levels of training may be appropriate based on whether an individual is directly or indirectly involved with the changes.

Review and Evaluation

Review and evaluation of a newly implemented control are crucial steps in the process of redesigning any problem job or task. Occasionally, the task redesign may overlook important, unintended consequences. Therefore, it is crucial to get feedback on how the changes are working and how well the miners are accepting the new procedures, as well as soliciting suggestions from them on how the new plan might be made more effective. Any safety concerns associated with the change should be vetted immediately. Long-term evaluation of effectiveness should consist of a systematic method of determining if the redesigned job or task has fulfilled the desired goal of reducing the worker's risk of injury. Such a long-term evaluation is necessary to determine whether the outcome meets the initial objectives of the project. It is important when comparing the new procedures to the old that the same data collection techniques be used so that an accurate comparison can be made regarding the effectiveness of the intervention.

It may be tempting to perform an evaluation immediately after implementation. However, it is important to allow enough time for the impact of the change to be fully realized. Inevitably, there are issues associated with the implementation during the break-in period for a new intervention. Evaluating the design changes before these issues have been fully integrated may give a false impression as to the effectiveness of the new procedures. Thus, it is best to wait until all of the "teething" problems have been addressed before evaluating the new system.

Facilities Design and Layout

Transportation of materials is costly in terms of space, machinery, and physical energy. It does not add value to the object being moved and exposes workers to numerous hazards. In fact, given that transportation costs for materials typically account for 30%–75% of the total operating cost, there is a strong economic incentive to improve the efficiency of materials-handling systems [Kroemer 1997]. However, some may view the reduction of LBP risk to the workers by redesigning, improving, or eliminating transportation barriers to be a more compelling motivation. Fortunately, ergonomic design of material-handling systems can benefit both the health of the worker and the company.

Efficient material flow is associated with few transportation moves, whether on the surface or underground. Analysis of current materials-handling practices is a critical step in the proper design of both existing and planned facilities. For existing facilities, it is often difficult to change the layout, but improvements in material flow can often be realized.

For existing facilities, generalized checklists have been developed that can help identify problem areas for typical materials-handling operations. Problem areas may include [Kulwiec 1985]:

- Crowded operating conditions
- Cluttered entries and supply areas
- Poor housekeeping
- Delays or backtracking in flow of material
- Obstacles in the flow of materials
- Manual handling of loads weighing more than 45 lb
- Excessive storage times for materials
- Single items being handled as opposed to unit loads
- Underutilization of materials-handling equipment where appropriate
- Excessive time required to retrieve stored parts or supplies
- Multiple handling of the same item

Videotape, still photography, portable tape recorders, and note-taking are some of the best methods to document the many elements of the job sequence. Videotape is the preferred recording system for a task analysis because the analyst can watch the job being performed in "real" time. Videotape can also be reviewed and analyzed as many times as needed to obtain information about the hazardous activities of a particular job. In addition, most videotape units allow the analyst to take verbal notes while videotaping. It is recommended that a minimum of three cycles of the task be recorded for analysis [Kirwan and Ainsworth 1992]. It may be necessary to record several job cycles if there is significant variability in how the task is performed.

It is often helpful to describe the flow of materials using a diagram or flowchart that shows the sequence and location of materials-handling activities or that represents a listing or table of steps associated with movement of a specific material [Kirwan and Ainsworth 1992]. This type of analysis can be very helpful in identifying unnecessary materials-handling activities and other inefficiencies associated with the supply-handling system. Obviously, such an analysis has the potential to reduce unnecessary manual lifting, which will in turn reduce the repetitive loading on the spine that leads to LBP.

The following general procedure of flow-charting has been recommended as a means of examining the efficiency of a supply-handling system [Kroemer 1997]:

1. Select the activity to be studied. Define this carefully so that your analysis will provide you with the answers you need.

2. Choose what you are going to follow (material or operator). A system can be examined in different ways. In some cases, it may be better to follow a specific material from delivery at the supply yard to its final end use. In other cases, it may be more useful to follow a specific worker who has responsibility for certain materials-handling activities. For example, it may be useful to follow a supply yard worker responsible for loading up a delivery of underground supplies to uncover inefficiencies or instances of manual lifting that might be eliminated.

3. Determine the starting and ending points. Decide on the limits of your analysis and do not exceed these when you perform the study.

4. Write a brief description of each detail in the flowchart. For each step, no matter how short, describe every operation, inspection, transport, storage, or delay encountered.

5. Apply the appropriate symbol. The following set of symbols has been suggested as a means of labeling activities observed in the analysis [Kroemer 1997]. They are by no means exhaustive, and others may be developed as needed:

OPERATION: An international change or action is being done (e.g., picking up an item, tightening a bolt, etc.

INSPECTION: When something is checked or verified but not changed, use a square to indicate an "inspection" step.

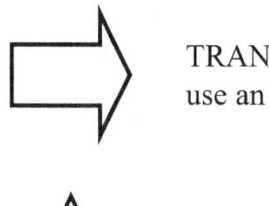
TRANSPORT: When a material is moved from one place to another, use an arrow to indicate "transportation."

STORAGE: When an item is purposefully put in place for a period of time, use a triangle to indicate "storage."

DELAY: Use a capital "D" to indicate when an object or person remains in one place, awaiting action.

6. Enter relevant information. Information such as distance traveled, time required, number of people involved, weight of objects, frequency of occurrence, etc., should be recorded, as necessary.

7. Describe hazards. These may include lifting demands, pinch points, sharp edges, or other hazards associated with different steps in the process.

8. Provide a summary. Summarize the hazards and statistics in the report and put into the summary block.

9. Provide comments. Provide your impression of the different steps in the process. Indicate problems/inefficiencies identified or the hazards you were most concerned about. Ideas on how the different steps in the process could be improved should be noted.

10. Control actions. In this step, questions need to be asked about each activity in order to help develop improvements to the design of the process. It is helpful to ask *WHO, WHAT, WHEN, WHERE, WHY,* and *HOW* questions:

- *WHO* performs the task? Are there enough people? Are these people the most appropriate? Can the job be made easier for them? Should they be wearing PPE?
- *WHAT* is being done? What is the purpose of doing it? Does the activity achieve the desired end? Can the activity be simplified in some way?
- *WHEN* is the activity being done? Why should it be done now?
- *WHERE* is the activity being done? Is this the best location to perform the activity? Where might be a better location to perform the task? How do the locations of preceding and subsequent tasks affect where this task is performed?
- *WHY* is this activity being done? Is it even necessary? Why is it done in this particular way? Can it be combined or reordered with other steps to make the process more efficient? How do preceding and subsequent tasks affect why and how this task is done?
- *HOW* is the activity being performed? Is the right equipment available? If not, why not? Are there better methods that could be used to accomplish this task?

11. Propose a better method. The preceding steps should make clear where the issues lie with the process, as well as what could be done to alleviate some of the existing problems. Write down your ideas on the chart, and propose the changes and their rationale to the appropriate personnel.

An example of such an analysis for two methods of installing roof screen in an underground coal mine is shown in Figure 20. The top chart in the figure represents the steps currently used to install the roof screen. The current method involves having a stack of roof screen delivered to the last open crosscut, where it is stored (resulting in a delay). This method requires miners to walk back to retrieve the screen (which takes up valuable time) and then manually lift and carry the screen over the bolter, causing a potential for back injury. The screen is then positioned on the automated temporary roof support (ATRS) and bolted into place. As can be seen, notations are made regarding the distances moved, time required, weight of objects, and frequencies per shift. In addition, hazards and other comments are recorded and control actions recommended.

The bottom chart in Figure 20 shows the proposed new method for installing roof screen. This entails installation of rails mounted on the bolter to facilitate storage and installation of the screen and a winch to help load the screen. In this scenario, the stack of screen is delivered to the last open crosscut and immediately loaded on the newly mounted rails using a winch. This eliminates the need for workers to walk back to the crosscut (saving time) and manually lift/ carry the screen up to the ATRS (eliminating manual lifting and back injury risk). Instead, the screen is moved forward as the bolter moves and the rails allow the miners to slide the screen rather than lift it, which is less physically demanding than lifting and carrying. The bottom chart in Figure 20 shows that the redesigned procedure is streamlined (requiring two fewer installation steps per cycle), time-saving, and has fewer hazards to the back.

As part of the strategy to eliminate or reduce manual handling of supplies and parts, the analyst should question all aspects of the supply-handling system. Suppose a videotape analysis shows that considerable disorganization exists at the surface supply yard at a mine. Questions to consider may include:

- What is the scope of the problem?
- How is it related to the entire system?
- Is it limited to the supply yard itself?
- Is it caused by inadequate on-site storage facilities, or is it caused by poor materials-handling practices?
- Could the problem be the way the materials are received from the suppliers?
- Can the schedule for delivery of supplies and equipment be more regimented instead of being dependent on the supplier's schedule?
- Can the supplies be received in a different configuration (already on a pallet or banded together instead of loose) to promote mechanical handling of materials?

The problem definition should contain quantitative information whenever possible. If there is a designated area for storage of parts or supplies, the following questions should be addressed:

- What are the dimensions of each storage compartment or area?
- How far away from the area of end use is it?
- How many different parts or supply items are stored there?
- How have materials been organized in the storage areas?
- Are they organized before they are delivered to the production sections?

An important aspect of the analysis process is encouraging the workers involved to provide observations and comments regarding potentially hazardous activities, risk factor exposure, or inefficiencies in the system. Workers may have already tried to incorporate some sort of change to the workplace or work method to alleviate the hazardous condition or risk factor exposures. Getting feedback from the workers regarding their suggested modifications sometimes provides the analyst with the beginnings of a permanent solution for those situations.

Method	Activity	Facts				Hazards					Comments	Control Action
Present ☒ Proposed ☐	Operation / Transport / Inspection / Delay / Storage	Distance (ft)	Time (min)	Weight (lbs)	Frequency (shift)	Falling Material	Hazard Material	Sharp Edges	Pinch Points	Bent Back		Eliminate Combine Redesign Simplify Reduce
1. Screen delivered to last open crosscut	○ ⇨ ☐ D △				5							
2. Storage in crosscut	○ ⇨ ☐ D △				5							Eliminate
3. Delay	○ ⇨ ☐ D △		30		5							Reduce
4. Miners walk back to retrieve screen	○ ⇨ ☐ D △	30	1		40							Eliminate
5. Miners lift screen	○ ⇨ ☐ D △			30	40			X	X	X	Back injury risk	Eliminate
6. Screen carried to bolter	○ ⇨ ☐ D △	30	1		40			X				Eliminate
7. Screen positioned on ATRS	○ ⇨ ☐ D △				40			X	X			
8. Screen bolted in place	○ ⇨ ☐ D △				40				X			
9. Repeat steps 3-8 eight times/cut	○ ⇨ ☐ D △											
10.	○ ⇨ ☐ D △											

Method	Activity	Facts				Hazards					Comments	Control Action
Present ☐ Proposed ☒	Operation / Transport / Inspection / Delay / Storage	Distance (ft)	Time (min)	Weight (lbs)	Frequency (shift)	Falling Material	Hazard Material	Sharp Edges	Pinch Points	Bent Back		Eliminate Combine Redesign Simplify PPE
1. Screen delivered to last open crosscut	○ ⇨ ☐ D △				5							
2. Screen loaded onto rails on bolter using winch	○ ⇨ ☐ D △	5			5							
3. Delay	○ ⇨ ☐ D △		22		8							
4. Screen slid up to ATRS using rails	○ ⇨ ☐ D △	10			40			X	X			
5. Screen positioned on ATRS	○ ⇨ ☐ D △				40			X	X			
6. Screen bolted in place	○ ⇨ ☐ D △				40				X			
7. Repeat steps 3-6 eight times/cut	○ ⇨ ☐ D △											
8.	○ ⇨ ☐ D △											
9.	○ ⇨ ☐ D △											
10.	○ ⇨ ☐ D △											

Figure 20.--Comparison of present and proposed methods for installing roof screen in an underground coal mine.

Use of Mechanical-assist Devices

It is often possible to develop or use mechanical-assist devices to prevent workers from having to perform forceful exertions. In many cases, existing materials-handling devices may be applied, particularly in surface operations, shops, prep plants, and other locations where headroom exists. In some environments, such as underground coal mining, there may not be existing mechanical devices that can be easily applied to solve a materials-handling problem. This section will briefly discuss general principles for the use of mechanical-assist devices and then present examples of effective interventions using mechanical-assist devices in different mine settings.

While use of mechanical-assist devices is one of the best methods of reducing risk of LBP from materials handling, care must be taken to implement such devices appropriately and effectively. One must consider certain aspects of the equipment itself, the environment in which it will be used, and proper training of the employees who will be using the equipment. It is critical to avoid introducing new hazards.

Mechanical-assist devices can take many forms, including simple tools, handcarts or dollies, conveyors, lifting devices, pallet trucks, and/or highly specialized tools built to address a unique situation. What these aids have in common is that all of them attempt to "lighten the load," eliminate the lift, and/or make materials handling more efficient. This can lead to higher productivity and a reduced risk of injury. Often the use of such devices can have such a positive impact on these factors that they can pay for themselves in a relatively short time.

While the benefits of mechanical aids can be quite dramatic, it is important to bear in mind that their use can also entail some risk if attention is not paid to proper implementation. It is important that the equipment being used for a particular load is of the appropriate capacity for the job. However, equipment with excessive capacity may also be a problem because higher forces may be required to move the device. Equipment used should be carefully matched and appropriate for the materials being handled, tasks being performed, and conditions under which it is being used. Ease of accessibility and ease of use are also critical to success, as workers are less likely to track down mechanical aids located in remote areas or to use devices that are time-consuming or difficult to operate. In addition, it is important to realize that many types of aids require regular inspection and maintenance for proper operation.

Proper implementation of mechanical aids also requires attention to the environment in which it will be used. If forklifts or other mobile equipment are used to assist with materials handling, it may be necessary to establish pedestrian walkways or block off an area during operation. Barriers should be placed to ensure employees are prevented from coming close to or positioning themselves underneath supported loads. Alarms or warning devices should be in proper working order and/or audible to persons working around the equipment. Passageways should be kept clear so that equipment can be easily maneuvered through these areas. If certain areas have space restrictions that make maneuvering more difficult, four swivel casters or wheels may be useful to increase maneuverability.

A final consideration that needs to be addressed is *proper training* for the employees who will be using the mechanical aids. All employees should be fully aware of safe operating procedures and appropriate work practices in the use of the equipment.

Application of Mechanical-assist Devices in Mining

While the mining industry has many unique environments where use of traditional mechanical aids is made difficult or impossible, there are usually many places on the mine property where traditional mechanical-assist devices can be used easily and effectively. The following examples illustrate some successful applications of the use of mechanical assists in mining.

Use of hoists for materials handling. One technique that has met with considerable success in the mining environment is the implementation of standard hoist mechanisms (both in the mine and on the surface) to assist with handling heavy parts or materials (Figure 21). Several mines have reported that installing hoists at central destination and delivery points can eliminate a significant amount of manual handling of heavy objects [Selan 1997].

Figure 21.—Hoists can be used at many locations on the surface and underground to lift heavy items.

Air bags. Air bags are remarkably versatile devices and are capable of lifting substantial amounts of weight. They are relatively lightweight and easy to transport when deflated. Figure 22 shows an air bag being used to lift a conveyor belt for belt roller maintenance. Some mines have used air bags to re-rail track vehicles that have become derailed underground. The air bag can be used to lift the derailed vehicle so that it can be easily repositioned over the rails. The bag can then be deflated so that the vehicle is properly positioned on the tracks.

Figure 22.—Using an air bag to lift a conveyor belt.

Mobile manipulator. In late 2006, the NIOSH Spokane Research Laboratory reported on the development of a prototype device designed for one person to safely and effectively lift and maneuver loads up to 600 lb for a variety of materials-handling tasks in maintenance shops and mines (Figure 23) [NIOSH 2006]. The device is mobile and has multiple speeds for tramming and turning. Some of the features of the "mobile manipulator" are the following:

- Compact and easy to stow or transport underground
- Full range of motion for ease of maneuvering objects
- Lifting pressure felt by operator limited to 10 lb
- Gripping attachments for specific tasks
- Braking on lifting arm for load control

Figure 23.—Mobile manipulator.

NIOSH has performed several field trials at sand and gravel shops and has recently signed a licensing agreement with a manufacturer of material-handling devices in an effort to put the mobile manipulator into industrial use.

Transfer of heavy items. Figure 24 shows excellent use of a mechanical assist to reduce the demands of a lifting task. The forklift is positioned to allow the item to be easily transferred from the storage rack to a pallet simply by rolling it onto the pallet. Note that a key to the effectiveness of this approach is that both the load storage and the forklift pallet are located at the worker's waist height. This minimizes the need to lift and allows the transfer to occur in an upright posture, both of which substantially reduce stress on the low back. It should be noted that an operator must be behind the wheel whenever forklifts are in a raised position, per Occupational Safety and Health Administration (OSHA) regulations.

Figure 24.—Using forklift and pallet for easy transfer of loads.

Clevis pin removal. Figure 25A illustrates the old method of removing a clevis pin during performance of maintenance activities. Obviously, removing the pin manually with a sledgehammer requires repetitive, high-force exertions that place a high demand on the back muscles. A new method was introduced (Figure 25B) that involves using a jackhammer end effector to accomplish the task and obviously eliminates the demands and low back stress associated with manual pin removal.

Figure 25.—Clevis pin removal: (A) manual removal with a sledgehammer; (B) removal using machine-mounted jackhammer.

Counterweighted lids to ease vault opening. There are many ways to reduce stress on the low back from common tasks. One example is shown in Figure 26. Opening the 100-lb metal vault lid shown in the figure was a stressful task and required significant application of force. Miners at the site devised the idea of installing hinges and attaching a counterweight to the large lid (seen in the left-hand side of Figure 26). This greatly reduced the force needed to open the lid, thereby decreasing the load on the low back.

Figure 26.—Counterweight installed to facilitate opening of vault.

Lift stands. Mine shops are areas where it is often easy to use off-the-shelf mechanical-assist devices, such as the lift stand shown in Figure 27. The task shown in the figure is the transfer of bar stock from the storage rack to a threading machine. Again, notice that the storage of the bar stock at waist level, in conjunction with the lift stand and adjustable shelf, allows easy transfer of heavy items with minimal lifting. Hoist, cranes, hand trucks, dollies, hand pallet trucks, and forklifts are mechanical-assist devices that are often very easy to use in shops or surface maintenance areas. Since maintenance tasks often require awkward postures, it is particularly important that mechanical aids be used wherever possible.

Figure 27.—Lift stands can have many uses in shop areas and can greatly facilitate load transfer.

Wheel chocks. Normal installation of wheel chocks involves bending down to place the chocks properly. However, one mine developed a simple intervention to reduce the amount of bending required to perform this task. Workers simply fastened a section of electrical conduit to the chocks, allowing the driver to position the chocks (and later remove them) without excessive bending of the trunk (Figure 28). When one considers the influence of trunk bending on low back loading (and the frequency with which wheels have to be chocked), even a simple intervention like this can significantly decrease the repetitive loading experienced by the low back and reduce risk of injury.

Figure 28.—Attaching a simple handle to wheel chocks greatly reduces excessive bending to place or remove the chocks [Torma-Krajewski et al. 2006].

Opening hopper car gates. Many mines ship their product via rail, and there are many stressful tasks associated with work in rail yards. One such task is opening the gates on a hopper car using a steel bar, as shown in Figure 29 (*left*). Using this technique, the worker has to adopt a flexed trunk posture and repeatedly pulls on the bar to crank open the gate. However, devices are available that can accomplish this task mechanically, as shown in Figure 29 (*right*). Use of the mechanical device eliminates a number of repeated and forceful exertions that would otherwise be necessary to open the gates manually (usually on multiple gates per car and on numerous cars during a shift).

Figure 29.—Using steel bar to open hopper gates *(left)*; gates opened via mechanical assist *(right)*.

Installing railcar shakers. Railcar shakers are used to dislodge material stuck in a hopper car during unloading. These shakers weigh approximately 50 lb, and placing them in the slots of the railcar involves awkward bending, reaching, and forceful exertions (Figure 30, *left*). This combination of weight and posture creates large compressive forces on the spine. However, there are simple devices that can ease the demands associated with this task. The device shown in Figure 30 (*right*) was fabricated in a local shop and allows easy handling of the shaker. This device allows the shaker to be wheeled to the hopper car and easily slid into place without any lifting or awkward postures. A number of different designs are available from rail safety equipment suppliers.

Figure 30.—Installing railcar shaker involves heavy load in flexed posture *(left)*; a simple device can be used to install shaker with a minimum of effort *(right)*.

Scoop-mounted forks and supply trays. Delivery of supplies in the underground environment can be made much more efficient if supplies are organized in trays (Figure 31, *left*) and then delivered via mechanical-assist devices such as scoop-mounted forks (Figure 31, *right*).

Figure 31.—Well-organized trays *(left)* can be efficiently transported underground using scoop-mounted forks *(right)*.

Winches. Mechanical aids such as winches can be used to improve handling practices for many items underground. Figure 32 shows a winch being used to load roof screen onto a bolter underground. Use of the winch eliminates the need for miners to manually lift and stack the screens on the bolter, saving wear and tear on the miners' backs.

Figure 32.—Use of a winch to facilitate loading roof screen.

Longwall supply transport. Another strategy that mines have had great success with is the development of vehicles to perform specialized functions. In many cases, such vehicles have been built entirely from salvaged parts and supplies, making these solutions quite cost-effective. Figure 33 shows a materials-handling cart called the Zipmobile (named after the miner who developed it). This cart rides on the handrails of the longwall conveyor and transports supplies along the longwall face. Instead of manually moving supplies beneath the longwall shields, miners can simply load up the cart and pull the supplies down the longwall face [Selan 1997].

Figure 33.—Cart that rides on conveyor handrails reduces the demands of transporting supplies down the longwall.

It is beyond the scope of this report to provide a comprehensive listing of all of the labor- and back-saving applications of mechanical-assist devices to reduce LBP risk in mining. However, this section has given some examples of applications that have either implemented or developed mechanical aids to reduce LBP risk in mining. There are scores of other examples.

Two useful resources regarding development and use of mechanical aids are: *Material Handling Devices for Underground Mines* [Conway and Unger 1989] and *Ergonomic Guidelines for Manual Material Handing* [NIOSH et al. 2007]. An Internet search can also be an effective method of exploring possible solutions. It is clear that mechanical-assist devices can be effectively used in both surface and underground environments. There is also no doubt that developing or implementing mechanical aids is one of the best strategies for reducing LBP risk in mining. Workers are often the best originators of labor- and back-saving devices and applications.

The Design of Lifting Tasks

When manual handling of materials cannot be eliminated through improved system design or the use of appropriate mechanical-assist devices, it is often possible to reduce the demands associated with manual lifting by improving the design of lifting tasks. Unfortunately, manual lifting tasks are often considered unavoidable and not enough thought is given to alternatives that can significantly decrease loads on the spine. However, there are some fairly straightforward design criteria that can dramatically reduce the loads experienced by workers during lifting tasks. This section provides some information on how to analyze lifting tasks and how to reduce the risks associated with lifting in mining environments.

Principles of Lifting

When one looks at the demands of lifting, it is obvious that the weight of objects being lifted is a major concern. It seems intuitive that lifting heavier objects would be associated with a higher risk of LBP, and research studies support this association [NRC 2001]. However, experts in LBP prevention know that the weight of an object is literally only half of the equation when assessing the stresses placed on the low back during lifting tasks. In fact, as will be demonstrated shortly, it is quite possible for back stresses to be higher when lifting a light versus a heavy object! To understand how this can be true requires an explanation of a fundamental principle critical to proper design of lifting tasks.

Consider a miner holding a crib block in two different orientations, as depicted in Figure 34. The crib block in this example is 6 in by 6 in by 30 in and weighs 25 lb. In Figure 34A, the miner holds the crib block with one end in each hand with the length of the block running across the front of the body. In Figure 34B, the miner is holding the block by one end with the length of the block sticking straight out from the body. You probably do not see many miners carrying crib blocks in this manner, and for good reason! Even though the crib block is the same weight in both cases, the miner in Figure 34B is experiencing much greater spinal stress compared to the miner in Figure 34A.

To understand the lifting principle involved here, the concept of the *center of gravity* of an object needs to be discussed. The center of gravity is an imaginary point where one can consider that the entire weight of an object is located (often this is in the center of the object, though not always). One can easily see that the center of gravity of the crib block is much closer to the miner's body in Figure 34A than in Figure 34B. This is important because the stresses imposed on the spine are a combination of the object weight and the distance (or lever arm) of the object's center of gravity from the spine. In fact, the effect of the object's weight needs to be *multiplied by this distance* to fully understand the load on the spine. In Figure 34A, the distance of the crib block's center of gravity to the spine is 1 ft. This means that holding the crib block in this manner creates a load of 25 lb by 1 ft, or 25 ft-lb. This represents a *torque* or a *moment of inertia* on the spine. In Figure 34B, the center of gravity is 2 ft from the spine, meaning that the torque (or "moment") on the spine is 25 lb by 2 ft, or 50 ft-lb. From this example, it can be seen that simply changing the orientation of the crib block can double the stresses experienced by the low back.

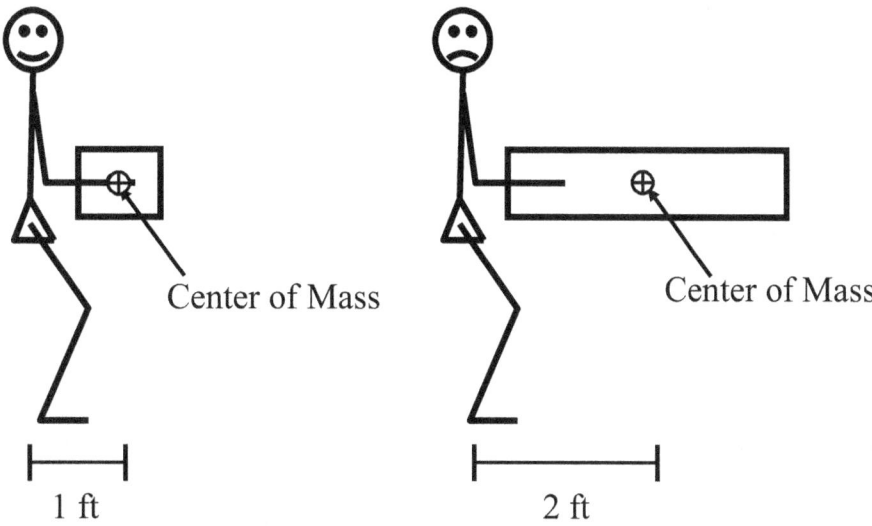

Figure 34.—Examples of holding a crib block. (A) With the crib block held by its sides, the distance from the spine to the center of gravity is 1 ft; (B) if held lengthwise, the center of gravity is 2 ft from the spine.

Understanding this concept is critical to the design of lifting tasks. In fact, the moment imposed by the load being lifted is one of the simplest and best predictors of the risk of low back disorders. Reducing the moment underlies many of the lifting recommendations of which you may already be aware. For example, one recommendation is to always "keep the load close." One can understand this advice when one realizes that increasing distance has a multiplying effect on the weight and on the stresses on the low back. There are two ways in which the moment can be reduced: reduce the weight of the object, or reduce the distance to the center of gravity of the load.

The Impact of the Load Moment on Back Muscle Force

A torque or a moment about the spine is actually a tendency to cause rotation. It can be seen from the previous example that gravity's effect on a load held in front of the body would have the tendency to cause the torso of the body to bend forward. For the torso to remain upright, there must be an equal and opposite moment acting in the opposite direction. In the case of the torso, the forces needed to generate this moment are supplied by the back muscles. However, while the lever arm for the load being handled can be as much as 40 in, the lever arm for the back muscles is only about 2 in! Because of the disparity in lever arms (distances from the spine) between the load and the back muscles, the back muscles often have to generate amazingly large forces in order to lift and/or hold heavy objects, as will be detailed in the following example.

Consider the situation depicted in Figure 35. A worker holds a 50-lb bag of cement. The center of gravity of the bag is 20 inches in front of the spine. This creates a 1,000 in-lb (forward-bending) moment about the spine. Back muscle forces must be used to counteract this forward-bending moment to keep the worker upright. Since the back muscles have a 2-in lever arm, one can calculate the necessary forces:

$$20 \text{ in} * 50 \text{ lb} = 2 \text{ in} * \text{lb of force from back muscles}$$

$$\frac{20 \text{ in} * 50 \text{ lb}}{2 \text{ in}} = \text{Force required from back muscles}$$

$$500 \text{ lb} = \text{Force required from back muscles}$$

(1)

So in order to balance this 50-lb bag of cement, the back muscles must generate 500 lb of force! The huge forces created by the contracting back muscles in such a case squeezes (or compresses) the spinal column. Such forces may be sufficient to damage some of the structures of the spine, specifically the vertebral endplate. It should be noted that the weight of the upper body (about 50% of body weight) would also be compressing the spine, adding to the overall compressive load. As discussed earlier in the section on "Causes of Low Back Pain," the endplates are the weakest link in the spine, and fractures in the endplate are believed to be an important factor in the development of disc degeneration and LBP.

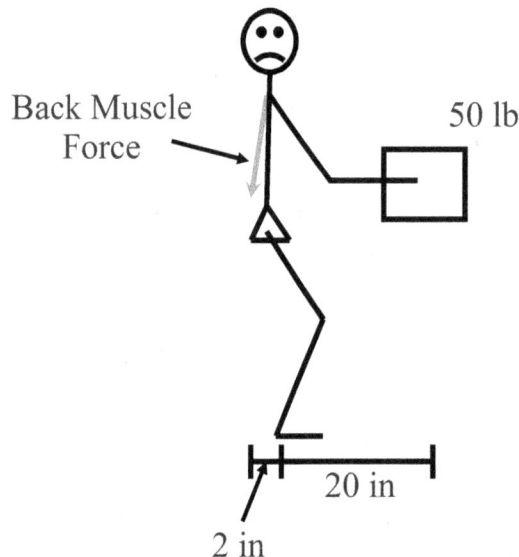

Figure 35.—In this example, the back muscles have a 2-in lever arm compared to the 20-in lever arm for the load. This means that back muscles must generate very high forces to balance the load.

Given the multiplying effect of distance on the weight of the object, it is possible to compare two lifting situations in which the lighter object actually creates more back stress than a heavier object! Such a situation is shown in Figure 36. Figure 36A shows a subject lifting a 40-lb object at 1 ft away from the spine. This lift creates a 40 ft-lb torque about the low back. By contrast, Figure 36B shows a worker lifting a 25-lb weight, but because a barrier prevents the worker from getting close to the object, the worker must lift it at a great distance from the spine (3 ft). The load moment calculation associated with this lift is 25 lb × 3 ft = 75 ft-lb, almost doubling the spinal stress compared to lifting the 40-lb load close to the body. In actuality, the situation in Figure 36B is even worse because the weight of the torso creates an additional forward bending moment, which adds an even greater demand on the back muscles and higher compressive forces on the back. The influence of torso posture will be examined in the next section.

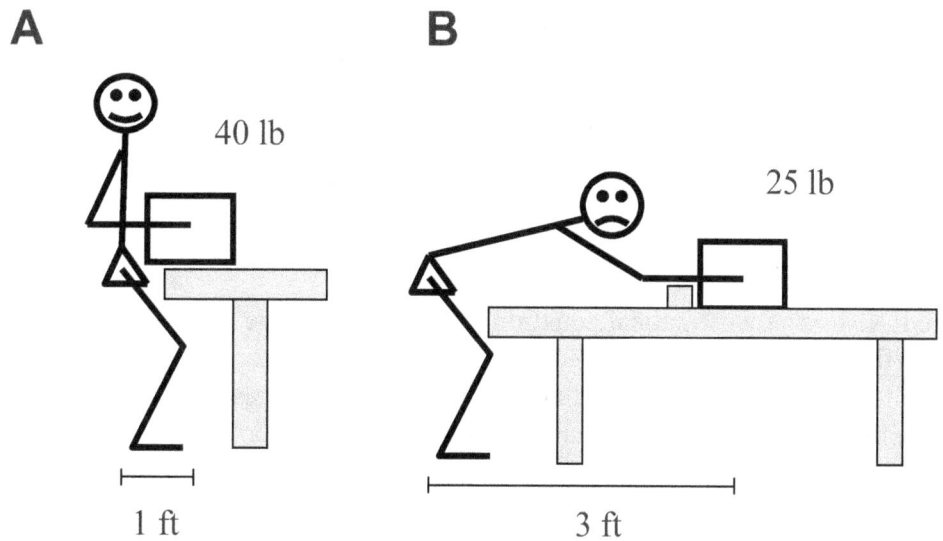

Figure 36.—Since weight of the load is multiplied by the distance away, a lighter object lifted at a great distance from the body (B) can actually create more spine stress than a heavy one lifted close to the body (A).

The Impact of Torso Flexion

The example shown in Figure 36 assumed (for ease of calculation) that the weight of the torso did not influence the load on the low back. However, bending the trunk complicates matters in several ways. The most obvious is that when the torso is bent forward, it creates an additional moment about the spine due to the force of gravity pulling on the weight of the head, arms, and torso of the body. Once again, a center of gravity for the torso, head, neck, and arms can be determined. Approximately half of the body's total weight is contained in these segments. The center of gravity is generally located about 8 in above the lowest lumbar disc. So for a 200-lb individual this creates another 135 ft-lb of forward-bending torque that must be counteracted by the back muscles, which will require an additional 540 lb of force from the back muscles. This will obviously create additional compressive stress on the spine and increases the likelihood of tissue failure of the spinal column.

Unfortunately, the floor (or ground) is a very common storage place for materials that are ultimately manually handled. Doing so greatly increases low back stress for the reasons just covered. An important lesson from this example is that items that must be manually handled should not be stored on the floor. Ideally, materials should be lifted from waist height when a standing posture can be adopted. The range of preferred starting points from which an item should be lifted is between knee and shoulder height. This is sometimes referred to as the "power zone," since more weight can be lifted safely within this region.

Improving the Design of Lifting Tasks

Based (in part) on the principles described above, NIOSH researchers developed a tool to help improve the design of lifting tasks. The most recent version of this tool is called the Revised NIOSH Lifting Equation (RNLE) [Waters et al. 1993]. Through a few relatively simple measurements and calculations, the RNLE allows an estimate of the amount of load that should be safe to lift for a large majority of workers. The following section briefly describes the RNLE. It should be noted that the RNLE is designed specifically for two-handed lifting tasks performed in an unrestricted standing position. It is not designed to address lifting tasks in postures involving prolonged stooping or kneeling postures, as are present in many underground coal mines. However, there are many jobs in the mining industry where the RNLE can and should be used.

The Revised NIOSH Lifting Equation (RNLE)

The RNLE is a mathematical equation for determining the recommended weight limit (RWL) and lifting index (LI) for selected two-handed manual lifting tasks. The RWL is the principal product of the RNLE. It is defined for a specific set of task conditions and represents the weight of the load that nearly all healthy workers could perform over a substantial period of time (e.g., up to 8 hr) without an increased risk of developing lifting-related LBP. "Healthy workers" refers to workers who are free of adverse health conditions that would increase their risk of musculoskeletal injury.

The concept behind the RNLE is to start with a recommended weight that is considered safe for an "ideal" lift (i.e., load constant equals 51 lb, or 23 kg) and then reduce the weight as the task becomes more stressful (i.e., as the task-related factors become less favorable). The RNLE consists of a fixed load constant of 51 lb that is reduced by six factors related to task geometry (i.e., location of the load relative to the worker at the initial lift-off and setdown points), task frequency and duration, and type of handhold on the object. The precise formulation of the RNLE for calculating the RWL is based on a multiplicative model that provides a weighting (i.e., multiplier) for each of six task variables:

- Horizontal distance of the load from the worker (H)
- Vertical height of the lift (V)
- Vertical displacement during the lift (D)
- Angle of asymmetry (A)
- Frequency (F) and duration of lifting
- Quality of the hand-to-object coupling (C)

The weightings are expressed as coefficients that serve to decrease the load constant, which represents the maximum RWL to be lifted under ideal conditions. For example, as the horizontal distance between the load and the worker increases, the RWL for that task would be reduced from the ideal starting weight.

The RWL is defined as follows:

$$RWL = LC \times HM \times VM \times DM \times AM \times FM \times CM \qquad (2)$$

where:

		Metric	U.S. Customary
LC =	Load constant =	23 kg	51 lb
HM =	Horizontal multiplier =	(25/H)	(10/H)
VM =	Vertical multiplier =	$1 - (.003\|V - 75\|)$	$1 - (.0075\|V - 30\|)$
DM =	Distance multiplier =	.82 + (4.5/D)	.82 + 1.8/D
AM =	Asymmetric multiplier =	$1 - (.0032A)$	
FM =	Frequency multiplier =	From Table A-1	
CM =	Coupling multiplier =	From Table A-2	
NOTE: Tables A-1 and A-2 are contained in the Appendix			

The term "task variables" refers to the measurable task-related measurements that are used as input data for the formula (i.e., H, V, D, A, F, C), whereas the term "multipliers" refers to the reduction coefficients in the equation (i.e., HM, VM, DM, AM, FM, CM). The following list briefly describes the measurements required to use the RNLE.

- H = Horizontal location of hands from midpoint between the inner ankle bones. This is measured, in centimeters or inches, at the origin and the destination of the lift.
- V = Vertical location of the hands from the floor. This is measured, in centimeters or inches, at the origin and destination of the lift.
- D = Vertical travel distance in centimeters or inches between the origin and the destination of the lift.
- A = Angle of asymmetry (angular displacement of the load from the worker's sagittal plane). This is measured in degrees at the origin and destination of the lift.
- F = Average frequency rate of lifting measured in lifts per minute.
  Duration is defined as: short-duration (≤1 hr), moderate-duration (>1 but ≤2 hr), or long-duration (>2 but ≤ 8 hr), assuming appropriate recovery allowances (see Table 1).
- C = Quality of hand-to-object coupling (quality of interface between the worker and the load being lifted). The quality of the coupling is categorized as good, fair, or poor, depending on the type and location of the coupling, the physical characteristics of the load, and the vertical height of the lift.

The LI provides a relative estimate of the level of physical stress associated with a particular manual lifting task. The estimate of the level of physical stress is defined by the relationship of the weight of the load lifted and the RWL. The LI is defined by the following equation:

$$LI = \frac{\text{Load Weight}}{\text{Recommended Weight Limit}} = \frac{L}{RWL} \qquad (3)$$

where load weight (L) = weight of the object lifted (pounds or kilograms).

According to Waters et al. [1993], the LI may be used to identify potentially hazardous lifting jobs or to compare the relative severity of two jobs for the purpose of evaluation and redesign. From their perspective, it is likely that lifting tasks with an LI > 1.0 pose an increased risk for lifting-related LBP for some fraction of the workforce. Lifting jobs should be designed to achieve an LI of 1.0 or less whenever possible. Some experts believe that worker selection criteria may be used to identify workers who can perform potentially stressful lifting tasks (i.e., lifting tasks that would exceed an LI of 1.0) without significantly increasing their risk of work-related injury above the baseline level. Those who endorse the use of selection criteria believe that the criteria must be based on research studies, empirical observations, or theoretical considerations that include job-related strength testing and/or aerobic capacity testing. Even these experts agree, however, that many workers will be at a significant risk of a work-related injury when performing highly stressful lifting tasks (i.e., lifting tasks that would exceed an LI of 3.0). "Informal" or "natural" selection of workers may occur in many jobs that require repetitive lifting tasks. According to some experts, this may result in a unique workforce that may be able to work above an LI of 1.0, at least in theory, without substantially increasing their risk of low back injuries above the baseline rate of injury. In one epidemiological study, however, NIOSH researchers demonstrated that risk of LBP for manual lifting increased as the LI increased, but was not significantly elevated until the LI exceeded a value of 2.0, i.e., the odds of workers in jobs with an LI value above 2.0 experiencing LBP was 2.5 times that for the unexposed control group [Waters et. al. 1999].

To provide a better understanding of the rationale for the development of the RWLs and LI, the article entitled "Revised NIOSH Equation for the Design and Evaluation of Manual Lifting Tasks" [Waters et al. 1993] discusses the criteria underlying the RNLE and the individual multipliers. The article also identifies both the assumptions and uncertainties in the scientific studies that associate manual lifting with low back injuries. For more detailed information about how to use the RNLE, one should consult the *Applications Manual for the Revised NIOSH Lifting Equation* [Waters et al. 1994], which is available for download from the NIOSH Web site (http://www.cdc.gov/niosh/docs/94-110).

The RNLE can be used as a tool to prioritize lifting tasks that need to be redesigned and can also be used on a "before and after" basis to evaluate whether changes made to a job have reduced LBP risk. If a lifting task cannot be made acceptable, other methods of task redesign should be considered (e.g., providing a mechanical-assist device).

Guidance for Design of Lifting Tasks in Restricted Postures

There are currently no specific tools (like the RNLE) to assess lifting tasks in kneeling postures or prolonged stooping postures, which are often used in underground coal mines. However, the results of studies on the physical capabilities and limitations of workers in these postures can be examined and can provide some general guidance for the design of lifting tasks in such postures.

Studies examining lifting capacity in the kneeling posture have shown a consistent decrease in lifting capacity compared to what can be lifted in a standing position. It seems that overall lifting strength is decreased from 7% to 21% in the kneeling posture (13% to 21% for repetitive lifting), probably due to the reduced leg strength available and perhaps the decreased balance and stability experienced in this posture [Ayoub et al. 1985a,b; Gallagher et al. 1988; Gallagher and Unger 1990; Gallagher and Hamrick 1992]. Lifting in a sitting posture results in a similar decrease in lifting capacity compared to kneeling [Ayoub et al. 1985a,b]. More significant decreases in lifting capacity (up to 33% less than standing) are seen when a squatting posture is maintained during lifting [Ayoub et al. 1985a,b].

The studies cited above make it very clear that lifting capacity is significantly reduced when a kneeling, sitting, or squatting posture is adopted. The implication is that manual lifting of supplies and other items in these postures should be designed to reflect this decreased capacity. In general, a 15%–20% decrease in weight would seem reasonable compared to what would be recommended in an unrestricted standing position.

Recommendations for design of lifting tasks when a stooping posture must be maintained are more complicated. This posture is associated with a number of problems that may lead to relatively rapid development of low back disorders. The stooping posture places very high stress on the low back muscles and ligaments and high compressive forces on the discs and endplates, even if no load is being lifted with the hands. When a load is lifted, these forces can quickly rise to levels where spine tissues get damaged fairly rapidly. What is perhaps most surprising is that when asked to assess the amount of weight they feel they can lift safely in a stooping posture, workers invariably select a weight that is similar to that in an unrestricted standing posture (and significantly higher than they select in a kneeling posture)!

This finding points out why the stooping posture poses a particular hazard for lifting tasks. This posture is a body stance where great strength is available to lift an object; however, the stresses on the low back are extremely high (approximately three times higher than those in an upright posture). It seems as though workers get feedback regarding the high-strength capabilities in this posture, but do not get sufficient feedback on the stresses experienced by the spine, which are high enough to cause rapid damage. Studies have shown that when the spine is bent forward (flexed), lifting a given weight can lead to failure of spine tissues three times faster than in an upright posture [Gallagher et al. 2005, 2007]. In addition, the stooping posture causes the ligaments of the spine to be stretched, which can actually lead to a loss of muscle strength and increased muscle dysfunction that can require significant recovery time.

Thus, while workers may feel capable of lifting a large amount of weight in a stooping posture, the biomechanical demands on the low back are such that spine tissue damage can occur fairly rapidly. Unfortunately, merely adopting the stooping posture (without lifting anything) can create a load on the spine that can cause tissue damage in some individuals. Lifting relatively light loads in this posture will lead to conditions where tissue damage is much more likely, and lifting heavy loads in a stooping posture (especially repetitively) will lead to conditions where tissue damage is very likely to occur in most individuals. This information should make it clear that the stooping posture is a very hazardous lifting posture to be avoided whenever possible. When the posture is unavoidable, weights that must be lifted should be kept to the minimum possible. Lifting of heavy weights should be extremely rare in this posture.

The essential point to consider in the design of lifting tasks in restricted postures is this: *restricted postures are all associated with a reduced capacity to lift weight*. In kneeling, sitting, and squatting postures, there is a reduced strength capacity and decreased stability when lifting. In the stooping posture, there is a need to limit the weight of loads due to the high stresses experienced by the low back. Operators of low-seam coal mines would do well to limit the weight of supply items and other frequently handled materials because of the physical limitations associated with work in restricted postures.

Other Tips for the Design of Lifting Tasks

Other changes can also be made to improve the design of lifting tasks. The following tips can be quite useful in reducing the demands associated with lifting tasks.

Tips for container design. Design of lifting tasks must also take into account certain aspects of the way in which materials are packaged. The following principles are among those that should be considered with respect to design of containers:

- Design containers so that mechanical handling is permitted.
- As a rule, it is better to lift lighter loads more often than heavier loads less often.
- Further reductions in load are necessary for items lifted in restricted postures (stooping or kneeling). In the kneeling posture, a 20% reduction in weight is recommended (<35 lb if possible). Loads lifted in a stooping posture should be as light as possible (<20 lb if possible).
- Package loads in the smallest container size possible.
- If packaging loose items (such as tools) in a container, package items to minimize shifting of items.
- Make sure load is evenly distributed.
- Design containers for lifting with both hands.
- Provide handgrips.
- Avoid sharp edges.

Workplace design and organization. Additionally, there are aspects of the workplace that should be considered and that can improve materials-handling practices. The following are some workplace design principles that can improve lifting tasks:

- If a standing posture can be used for lifting, store items between knee and shoulder height (not on the floor). Some refer to this as the "power zone" for lifting tasks.
- Floor storage (if unavoidable) should be reserved for small or light items.
- Eliminate barriers that prevent getting close to the load.
- Organize lifts so that distances that items are lifted or carried are minimized.
- Try to organize work so that physical demands gradually increase as the day progresses.
- If possible, alternate lifting tasks with nonlifting tasks during the workday.
- If possible, alternate lifting tasks among workers to distribute the load.
- Avoid prolonged periods of lifting, and make sure to take adequate breaks.
- Maintain good floor conditions.

Lifting tips. When lifting is performed, attention to the following items may be helpful [NIOSH et al. 2007]:

- Always keep the load as close as possible to the body.
- Avoid forward bending, especially in the first hour of a shift, when risks are greater.
- Avoid twisting the trunk.
- Use both hands if possible.
- Stretch your muscles periodically.
- Make sure you have sturdy footwear with nonslip treads.
- Make sure you have a firm footing.
- Test the load for stability and weight.

- Get a secure grip. Note that while gloves can be helpful when carrying objects with sharp edges, they can also reduce gripping strength up to 40%.
- Use extra caution if loads are unstable.
- Slide loads whenever possible (instead of lifting).
- If lifting a long object (such as a timber), support one end and just lift the other. Supporting loads on a surface will decrease the forces needed for the lift.
- Team lifting should be used for awkward or heavy items.
- Don't toss items that weigh more than 5 lb.

Principles of Seating and Control of Whole-Body Vibration

Many jobs in the mining industry require operators to drive vehicles or equipment in a seated posture with exposure to WBV. Many people do not realize it, but in the seated posture the spine is flexed forward almost to the point that it would be in a full stoop if one were standing on one's feet. When vibration, jolts, and jarring are applied to the flexed spine, it is a recipe for development of LBP. One can easily visualize that repeated exposure to the types of vibrations, jolts, and jarring experienced when driving a haul truck over a rough road might lead to fatigue failure of the vertebral endplates, just as loads with repeated lifting can. Fortunately, there are ways to decrease this risk by proper design of seating and by instituting control measures to reduce WBV.

Seat Design: General Considerations

According to Pheasant [1986], the purpose of a seat is to provide stable support to the body in a posture that is (1) comfortable over a period of time, (2) physiologically satisfactory (e.g., does not cut off blood flow or nerve conduction to any part of the body), and (3) appropriate to the task or activity at hand. Comfort is an individual perception and varies between people. Usually, a comfortable seat will also be physiologically satisfactory; however, a comfortable seat may not always be the most appropriate for the task. A recliner is a comfortable seat, for example, but is not appropriate for working at an office desk. Fundamental criteria for seat design will be discussed first, followed by design of seats for mobile equipment.

Given the wide range of sizes of different individuals, it is clear that a seat used by more than one person will need to be adjustable so that different sizes of workers can be properly accommodated. Workers must be trained (and encouraged) to make proper seat adjustments, and the vehicle should be designed and maintained to facilitate ease of adjustment. In this analysis, primary consideration will be given to addressing the effects of seating on the low back. However, other aspects will also need to be addressed since there are connections between various components of the chair and the body.

If one considers sitting on a flat surface (such as a table) with the feet dangling and no support from the arms, the entire body weight is supported by two bones (the ischial tuberosities) and a small portion of the thighs. The typical position for the back in this type of unsupported sitting is a slumped posture (i.e., the spine is severely bent forward). A properly designed seat will provide a good backrest support to the torso so that an upright (not slumped) posture can be maintained. Slumped sitting increases stress on the low back and needs to be avoided. A slightly reclined backrest with a 100°–110° angle between the thigh and the torso greatly reduces low back stress. Therefore, key considerations in seat design are to provide a slightly reclined backrest, to increase the thigh-torso angle, and to try to afford a slight hollowed-out curve in the spine to place the spine in a more stable and less stressful posture. The following sections discuss various chair components and how these may be optimized to reduce stress on the spine when seated.

The Backrest and Lumbar Support

In general, higher backrests are favored because they can effectively support more of the body weight and distribute the load better. However, there are some circumstances in which a lower backrest should be used in order to allow movement of the shoulders. This is often required in mobile equipment when operators must turn to look behind them in vehicle operation. If a high-level backrest is possible, the backrest height should be 35 in to accommodate a large male, measured from the back end of the seat pan. Medium-level backrests need to be 25 in high for proper shoulder support. Low-level backrests should be at least 15 in high [Pheasant 1986]. Low back stress is reduced the farther back one is reclined; however, a compromise usually must be made so that a seated worker can see what he or she is working on or where he or she is going. The seat pan-backrest should be greater than 90°, and if forward visual attention is required, a backrest of up to 105° can be maintained. As mentioned previously, backrests that are more reclined than this will reduce stress on the low back, but may inhibit the ability to see what is going on in front of the body.

Seats should provide lumbar support to maintain (as far as possible) a hollow in the low back region. This is necessary to maintain a proper curve in the lumbar spine. The depth of the lumbar support curve should be approximately 0.6–0.8 in from front to back and the maximum convexity should occur approximately 9 in above the seat pan (Figure 37). It is preferable that the degree of lumbar support be adjustable.

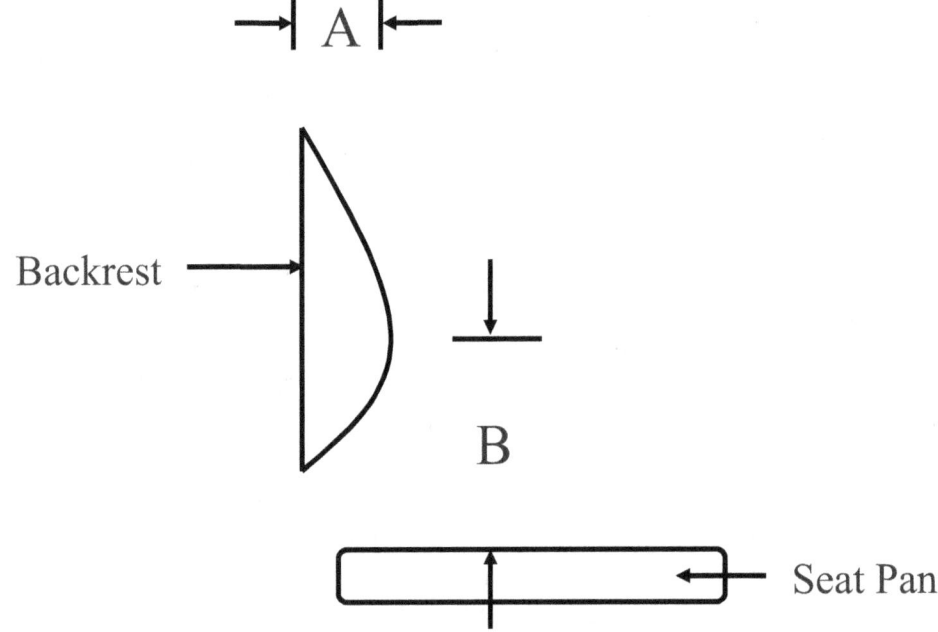

Figure 37.—Suggested dimensions for design of lumbar support. Maximum depth of the lumbar support (A) should be 0.6-0.8 in. The maximum convexity (B) should occur at a height of 9 in. above the seat pan. The seat should allow adjustment of these dimensions for proper fit.

Seat Pan

The depth of the seat pan should be 15–17 in, and its width should be approximately 18 in, according to the ANSI/HFES 100 standard [ANSI-HFES 1988]. The height of the seat pan should be adjustable and should be approximately 15–20 in above floor level to fit the majority of workers. Tilting the seat pan backwards slightly can help the worker maintain contact with the backrest.

Armrests

Armrests are an individual preference. They can provide added postural support in a seated posture and can be an aid to workers with respect to standing and sitting motions. It is best if the height of the armrests is adjustable so that they can be tailored to the individual using them. Fixed armrests can result in awkward forearm postures and a lack of proper support. If not well designed, armrests can prevent users from being able to reach work surfaces, objects, or controls. In addition, a poorly designed armrest can snag clothing, which may pose a significant hazard. Armrests should provide padding, be contoured to avoid sharp edges, and provide a durable, smooth surface on which to rest the arms. Use of armrests can lead to an improved distribution of body weight and decrease low back load.

Seating Design for Mobile Equipment

The design of seats in mobile equipment should generally comply with the recommendations given above to provide proper support to the body and a good posture from which to work, but may also have to provide some protection against WBV. As mentioned earlier, there is strong evidence that exposure to WBV is associated with the development of LBP. Mine workers can be exposed to WBV in both surface and underground operations, and the control measures may vary significantly depending on where (and in what type of equipment) the worker is exposed. The following sections briefly discuss issues of seating and WBV control in surface and underground operations. A thorough treatment of these topics is beyond the scope of this report, but sources for additional information are cited throughout this section.

Surface mining. One should be aware that there are international standards available as a design guideline for WBV exposure in vehicles. Specifically, the International Organization for Standardization (ISO) has recommendations that should be followed to avoid undue exposure to WBV [ISO 1997]. Unfortunately, studies of occupational drivers indicate that these standards are often exceeded in practice.

As mentioned earlier, the jolting and jarring present when exposed to WBV can substantially increase spine loading and may lead to fatigue failure of spinal tissues if exposed at a high enough level or for a long enough time. Unfortunately, severe jolting and jarring may cause damage to spinal tissues, leading to LBP [Bovenzi and Hulshof 1999]. One way of reducing the load on the spine is to avoid vibrating the spine at its resonant frequency, where the effect of the vibration can be greatly magnified. In the case of the seated spine, the primary resonant frequency is within a band of approximately 4.5–5.5 Hz. Secondary resonant frequencies can be found at the 9.4- to 13.1-Hz range. It is possible to damp vibration energy in these ranges, and doing so will greatly decrease the load experienced by the spine. The primary resonant frequency (around 5 Hz) has also been shown to increase the rate of fatigue of the back muscles. After exposure to WBV, the back muscles are not able to contract as quickly. This suggests that unloading a vehicle after a period of WBV may pose an increased risk to the back as a result of the muscle fatigue.

In general, the following recommendations can be given to reduce the risk of LBP associated with exposure to WBV on surface equipment. First, WBV exposure should be controlled using suspension systems and/or attenuated via cab or seat suspensions to comply with ISO standards [ISO 1997]. It should be noted that installing a new seat does not necessarily reduce WBV exposure. In fact, measurements of new seats have sometimes shown that WBV is increased! Control of WBV is complex and involves numerous factors (such as vehicle speed, road conditions, and the amount of load being hauled). Professional evaluation and analysis of vibration exposure are necessary to ensure that WBV has been properly controlled.

Other design recommendations include ensuring that cab and seat design is appropriate so that awkward postures can be avoided. Extended shifts are discouraged when significant WBV is present, and frequent stretch breaks should be allowed. Proper maintenance of roadways can also help reduce WBV exposure. Lifting immediately after extended periods of WBV should be avoided if possible.

Underground mining. Development of ergonomically designed operator compartments and work stations for underground mining equipment is an imposing task. The interaction of the confined space of the mine and the massive equipment required to mine the coal often results in cramped and poorly designed operator compartments. It is common to find operator compartments less than 30 in high and less than 24 in wide. A study of mobile underground equipment operators in the early 1980s indicated that between 33%–39% of operators were exposed to levels of WBV exceeding the ISO standards set to maintain performance and prevent fatigue [Remington et al. 1984]. Between 7% and 14% were exposed to levels exceeding the ISO maximum exposure limit. Other tests have been reported that miners may experience nearly 35% of the ISO 8-hr exposure limit simply riding for the 30 min it takes to reach the working face of the mine [Love et al. 1992].

The goals for seating design in underground mobile equipment are similar to those in other applications: provide a stable position from which to control the machine, provide some isolation from vibration and jolting, and reduce the risk of postural fatigue [McPhee 1993; Collier et al. 1986]. Guidelines have been put forth for the design of seating in underground equipment for various seam heights [Collier et al. 1986; Mason 1992; Canyon Research Group 1982]. Figure 38 illustrates the effects of vertical space restrictions on seating envelopes for the 95th-percentile miner for two cabs, 42 and 22 in high [Canyon Research Group 1982]. Comparison of the two cabs clearly illustrates the increased cab length, reduced reach envelopes, and restricted field of vision associated with a reclined seating posture.

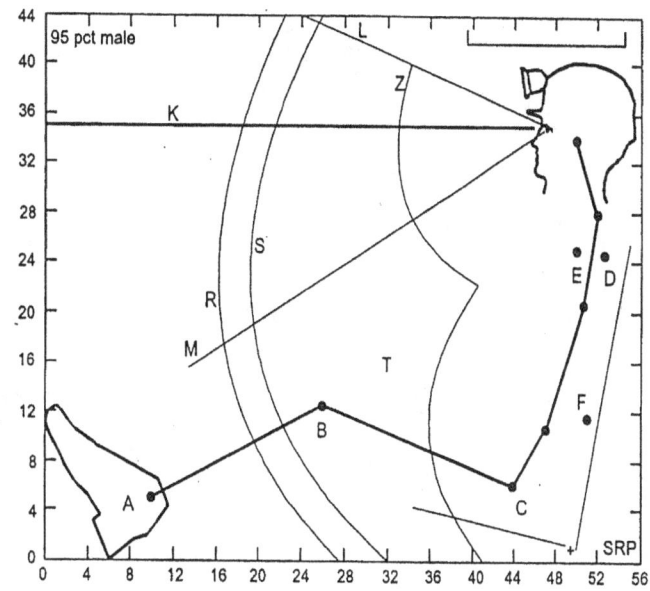

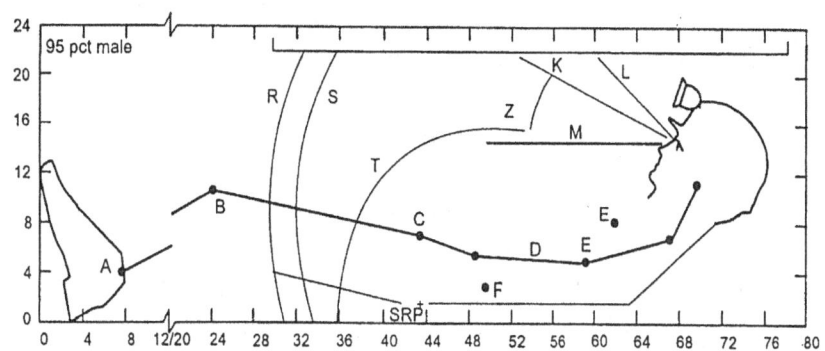

Figure 38.—Seating space envelopes for 95th-percentile miners in 42-in *(top)* and 22-in *(bottom)* compartments [Canyon Research Group 1982].

Collier et al. [1986] presented three design options for underground equipment seating. These included "normal" seating (for canopies 58 in high), a version with an increased backrest angle for canopies between 56.7 and 57.5 in, and a "constant eye height" option for canopies between 45 and 51 in. A special consideration for mining equipment seating is the provision of sufficient space so that the operator's cap lamp battery and self-contained self-rescuer can be worn on the belt. An example of such a design is shown in Figure 39 [Mayton et al. 2006].

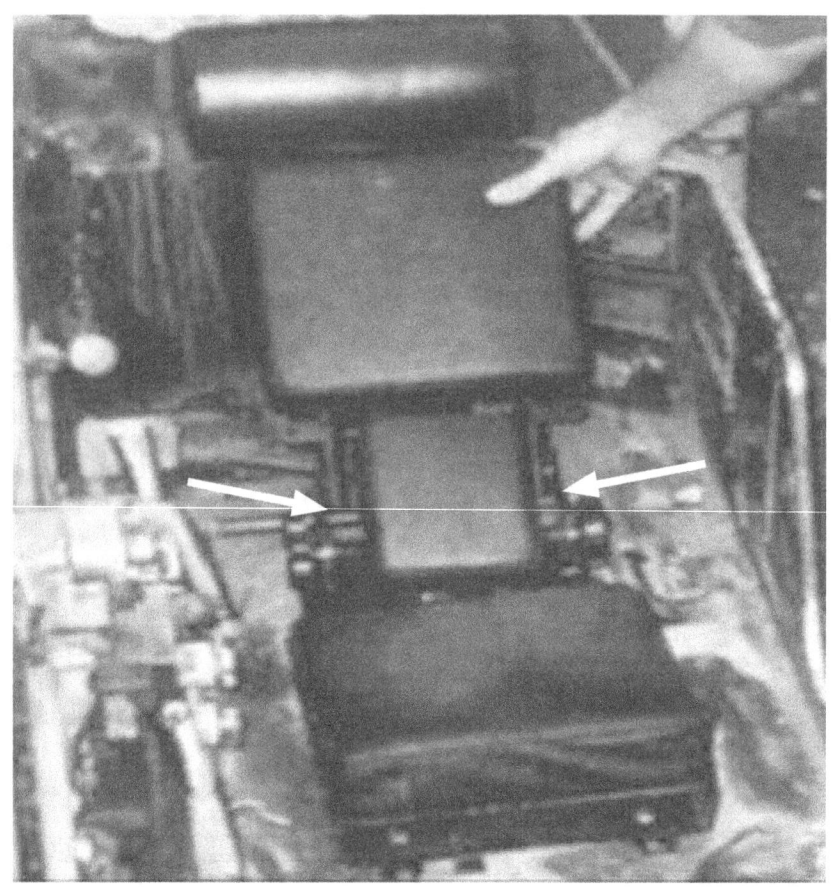

Figure 39.—Prototype seat design for underground mining equipment. Note cutouts in the seat backs (indicated by arrows) to accommodate belt-worn PPE [Mayton et al. 2006].

Mayton et al. [1999, 2005, 2006] present greater detail on methods of providing WBV isolation using viscoelastic foams in the cushions of seats on underground mobile equipment. These foams afford a significant decrease in the vibration energy transmitted to the seated operator and are currently being offered by some original equipment manufacturers as a seating option for underground equipment [Mayton et al. 2005].

Psychosocial Factors, Work Stress, and Low Back Pain

Psychosocial job stresses, such as job dissatisfaction, monotony, and time stress have demonstrated clear relationships to LBP incidence rates [NRC 2001]. An optimal psychosocial work environment for workers is characterized by demands that are adapted to an individual's capacities (psychological demands), a satisfactory level of influence (decision latitude), adequate social support from superiors and colleagues, a balance between efforts expended at work and reward received, predictability of work, and meaning of work [Karasek and Theorell 1990; Siegrist 1996; Kristensen 1999].

According to Elkin and Rosch [1990], work redesign and organizational development have been considered effective measures for reducing psychosocial stress. Organizational strategies for reducing psychosocial stress include the redesign of activities, participative and supportive management, career development, establishment of goals and cohesive teams, and flexible employee policies. Job rotation and other methods of enlarging the employee's job responsibilities can help improve job satisfaction. Lack of job control is perhaps the most critical psychosocial work factor [Sauter et al. 1989]. Improved job control can be achieved by allowing workers to have input into decisions that affect their job and by increasing worker participation in the production process. In organizational development and job redesign, employees' participation in planning, implementing, and evaluating work changes is crucial [Karasek and Theorell 1990]. Studies on stress at work have found different approaches inside organizations, but general themes involve establishing participatory systems and increasing individual control over one's own situation [ILO 1992; Lindström 1995]. Some mines have successfully employed the autonomous work group approach, which can be an effective method of increasing worker control and enriching jobs [Torma-Krajewski and Lehman 2007]. This team approach can also have benefits in terms of improved socialization and learning opportunities [Carayon and Lim 2006].

It is understandable that workers would desire increased control, decision latitude, and improved job satisfaction. However, how do these relate to LBP rates? There may be several explanations. One is that stresses such as time pressure can change work behaviors, even to the point of exerting more force than necessary or rushing excessively and experiencing an injury as a result. Psychological stress has also been found to increase muscle tension and the circulation of stress hormones in the body, which may influence pain perception [NRC 2001]. However, the issue of how psychosocial factors influence reporting of LBP may be more psychological than physiological. If a worker is currently dissatisfied with his or her job, even a relatively minor bout of LBP may provide the worker with a means to get out (at least temporarily) of what he or she perceives as a negative work situation. On the other hand, a worker who has been given increased responsibility and decision latitude by the company is more likely to embody the attitude that "I am valued by my company and they need me at work." Such a worker may be more willing to work through minor episodes of LBP. There is strong evidence that psychological factors such as these are associated with the reporting of LBP [Snook 2004a]. Importantly, a review of the literature by Linton [2000] concluded that "there is strong evidence that psychosocial variables are strongly linked to the transition from acute to chronic pain disability," i.e., psychosocial factors may play an important role in the expansion of LBP incidents to cases that are more costly and time-consuming.

Planning a psychosocial intervention in a real organization is a complex process, and during the intervention many unexpected changes can occur. Mines can change ownership; go into bankruptcy; change structure, managers, and employees; or even change mining technology in a short period of time. However, benefits are seen with these types of interventions including lower absenteeism [Lavoie-Tremblay et al. 2005]. It may never be possible to achieve a job that does not have some negative psychosocial stresses; however, it is clear that negative psychosocial work factors need to be eliminated to the degree possible. In some cases, it may be possible to counteract some negative psychosocial stresses that are inherent in a job by adding some positive psychosocial work factors, such as increased decision latitude [Carayon and Lim 2006]. One thing is clear: psychosocial factors play an important role in the control of LBP cases, and any program that seeks to have a comprehensive approach to LBP control needs to pay attention to this issue and reduce negative psychosocial work factors at the mine site. Occupational health psychologists or others can help evaluate and provide advice regarding psychosocial interventions.

IV. SECONDARY PREVENTION: REDUCING DISABILITY

The primary prevention approaches mentioned above have been shown to be effective in reducing the incidence and cost of LBP in numerous occupational settings, and every effort should be made to use the techniques described in the previous section to prevent LBP to the fullest extent possible. Primary prevention methods are clearly preferred, as interventions that prevent the initial injury avoid a multitude of subsequent problems that can be very costly. However, while primary prevention should be the main focus of LBP prevention efforts, it must be recognized that not all LBP is preventable at the current time. Thus, it is important for any program that wishes to control the impact of LBP in the workplace to address the issue of disability (i.e., lost time from work) caused by LBP. Fortunately, there are many techniques that have proven effective in reducing disability, and these seem to have substantial benefits for everyone affected. Specifically, Snook [2004b, 2006] suggests that the following approaches can lead to effective control of low back disability: workplace design, proactive return-to-work programs, communication, and management commitment.

Workplace Design

Workplace design in the context of primary prevention (prevention of initial episodes) of LBP has already been discussed. However, workplaces need to be designed to accommodate workers with LBP as well as those without. In fact, good workplace design can facilitate the return of injured workers. This is essential in maintaining a positive relationship between the injured worker and the company. In some cases, proper design of the workplace will also allow employees with LBP to remain on the job rather than missing work. It is important to note that workplace design may also be a benefit in preventing the recurrence of injury.

Workers with LBP often have difficulty tolerating awkward postures. Typically, these include forward bending and prolonged sitting. Unfortunately, in some mining environments (e.g., in low coal or during field maintenance activities), reducing the amount of forward bending is difficult, if not impossible, to achieve due to the environmental restrictions imposed on the worker. In some cases, it may be necessary to provide alternative work. However, it is often possible to reduce bending in other underground environments and/or prep plants, shops, and other mining facilities. Excessive bending is detrimental in many respects. It is time-consuming, increases the probability of low back disability, and increases the chance of aggravating an existing disorder. If anything, avoiding flexion is even more important for the injured worker.

Depending on the location, jobs can be designed to accommodate injured workers by providing improved seating, standing work stations, or providing materials-handling aids. One study found that use of lift tables and lifting aids (e.g., overhead pulley systems, vacuum hoists) significantly reduced the low back disorder rate in repetitive manual handling jobs [Marras et al. 2000]. As also discussed previously, one of the best ways to avoid torso flexion in unrestricted settings is not storing materials on the floor. Workers with LBP also have difficulty handling objects when excessive forward reaching is involved [Marras et al. 1995; Zurada et al. 2004]. Objects can often be handled with little stress if excessive bending and forward reaching are avoided. As also discussed previously, many mechanical aids and design strategies are available that can reduce the stresses of bending and reaching in the workplace. Miners experiencing LBP who must function in low-coal mines should particularly avoid the stooping posture to the greatest degree possible, especially when handling loads. Manual handling guidelines for workers with LBP have been developed by Ferguson et al. [2005].

Workers with LBP may also have difficulty sitting for long periods of time. It may be possible to provide a work station where these workers can alternatively sit or stand in order to change their posture frequently and at their own discretion to minimize any pain or discomfort.

Results from several studies indicate that making accommodations in the workplace through job design can substantially reduce lost time from work resulting from LBP cases. A study of 130 workers who had been off work for 6 weeks because of LBP investigated the effects of an intervention involving workplace redesign and time-limited light duty. The study found that the time to get workers back on the job was reduced from 131 to 67 days [Loisel et al. 1997]. A prospective study of one company examined changing from standard care to an approach that included worker rotation schedules, reduced lifting loads, and ergonomic redesign of tasks. The total days lost per 100,000 hr worked dropped from 60.9 days for standard care to 1.1 days for the occupational management approach, and then increased again to 95.8 days when the company selected an early intervention program of intensive physical therapy and work hardening [Lemstra and Olszynski 2003, 2004]. Management is sometimes reluctant to pay for workplace redesign due to the up-front costs involved. However, such an investment can pay substantial dividends in terms of reduced compensation costs and increased worker performance. An economic justification in terms of the "payback period" may be necessary to convince management of the cost-effectiveness of this approach [Seeley and Marklin 2003].

Proactive Return-to-Work Program

Data from several studies have shown that the longer a worker is off from work because of a back injury, the less likely it is that the worker will return to productive employment [Andersson et al. 1983; Carter and Birrell 2000; Hashemi et al. 1997]. Once a worker is off work for 4–12 weeks, he or she has a 10%–40% risk (depending on the workplace setting) of still being off work at one year. After a 1–2 year absence, it is unlikely that he or she will return to any form of work [Carter and Birrell 2000]. These studies underscore the importance of getting workers back to the job as quickly as possible by providing modified, alternative, or part-time work to the injured employee. Data indicate that an early return to work is in the best interests of everyone: the worker, the company, and, if appropriate, the union [Snook 2004b, 2006].

These statistics emphasize the importance of providing modified, alternative, or part-time work as a means of returning the disabled employee to the job as quickly as possible. The U.K. occupational health guidelines recommend that the worker be encouraged to remain in his or her job or to return at an early stage, even if there is still some LBP [Carter and Birrell 2000]. There is moderate evidence that the temporary provision of lighter or modified duties facilitates return to work and reduces time off work [Carter and Birrell 2000]. A review of the literature by Krause et al. [1998] concluded that modified work programs reduce the number of lost workdays by approximately half. However, a review of return-to-work interventions by Staal et al. [2002] revealed that ergonomic measures were applied in only 3 of 19 interventions.

A proactive return-to-work program is a supportive, company-based intervention for personally assisting the disabled employee from the beginning of the episode to its positive resolution. In a proactive program, the actions and responsibilities of individuals within the company and external providers are spelled out and related to the goal of resumption of employment [Hunt and Habeck 1993]. In the Michigan Disability Prevention Study, companies reporting 10% greater level of achievement on the proactive return-to-work program variable demonstrated a 13.6% lower rate of lost workday cases [Hunt and Habeck 1993; Hunt et al. 1993]. Mining operations remain a difficult environment in which to get injured workers back to a "return-to-work" level due to the unique environment and physical demands present in many jobs. However, some mines have been able to come up with alternative work that can help the employee return to work.

Communication

There is evidence that communication, cooperation, and common agreed goals among the worker with LBP, the occupational health team, supervisors, management, and primary health care professionals are fundamental for improving clinical and occupational health management and outcomes related to LBP [Carter and Birrell 2000]. When workers become temporarily disabled, it is important that management establish and maintain good communications with the worker and appropriate medical personnel. Supervisors should be instructed to follow up every disability case with a telephone call or visit before 2 days of lost work time have elapsed. The purpose of the call is to let the worker know that the company is concerned and to have the worker inform the supervisor of the status of his or her recovery. One company recently instituted a program that increased the communication among the worker, employer, medical practitioner, and insurer. When a worker compensation claim was received, the employer made immediate contact with the worker and insurer and followed up with calls at regular 10-day intervals to make certain that the claim was progressing smoothly. The tone of the communication was always pleasant, and the focus was always on the best interests of the worker. The message strongly communicated to the worker was: "You are a vital part of our team, your work is important, and your job is waiting for you." The program reduced the proportion of long-term low back disability claims from 7.1% to 1.7%. The possibility of retraining was explored for extended claims, and a liaison was established between management and the insurer if a gradual return to work was indicated. The focus of all communications was that every action taken was in the best interest of the worker. This program significantly reduced the proportion of long-term worker compensation claims and also significantly reversed a trend of increasing accident rates [Snook 2006]. The essential point is that positive communication with injured employees lets them know that both they and their contributions are valued by management. This cannot help but foster improved morale and increased loyalty of the workers who have been shown such concern.

Management Commitment

Management can sometimes complicate matters by not responding well to LBP when it does occur among workers. The Michigan Disability Prevention Study demonstrated that lower levels of disability are associated with management policies and practices, particularly safety diligence, safety training, and proactive return-to-work programs [Hunt and Habeck 1993]. Management plays a key role in keeping workers on the job and in returning them to work quickly. Goals and objectives and standard operating procedures (SOPs) should be provided by management, along with proper resources, to foster the safety and health program and return to work for injured employees.

Occupational safety and health literature stresses the importance of management commitment in successful safety and health programs [Griffiths 1985; OSHA 1990; GAO 1997; Gaspers 2002]. Safety must be managed in the same way as other high-priority company programs, e.g., finance, sales, production, and advertising. One researcher asserts that the chief executive officer must also become the chief safety officer [Griffiths 1985]. Specific ways in which management commitment can be demonstrated include establishing goals, assigning staff, providing staff time, making resources available, evaluating results, communicating with employees, and encouraging employee involvement in the program. The intent is to create a corporate culture with a positive and supportive attitude toward employees with LBP (or other health-related issues). Trust building and employee advocacy are important ingredients.

According to the U.K. occupational health guidelines, there is general consensus but limited scientific evidence that workplace organizational and/or management strategies may reduce absenteeism and duration of work loss [Carter and Birrell 2000]. One study reported a decrease in lost-time LBP cases from 20 to 2 per year after an improvement in top management commitment [Griffiths 1985]. High job satisfaction and good industrial relations are the most important psychosocial characteristics associated with low disability and sickness absence rates attributed to LBP [Carter and Birrell 2000].

On occasion, workers experience indifference or hostility from supervisors after reporting LBP. Adversary situations are set up, and employees look for ways to retaliate. Prolonged disability is often the result, with increased costs for both management and the employee. Supervisors must be trained in the true nature of nonspecific LBP, i.e., it is a disorder of unknown cause that happens to practically everyone, usually develops gradually, may recur frequently, and does not respond well to treatment but usually resolves itself within a few days or weeks [Snook 2004b].

Wellness Behaviors During Low Back Pain Episodes

As mentioned previously, most patients with LBP (up to 85%) cannot be given a precise diagnosis as to the cause. In most cases, however, the cause does not involve a serious systemic disease. Once a systemic disease has been ruled out, what course of action should be taken? Depending on the type of practitioner visited, several treatment options may be attempted [Nachemson 2000]. However, despite the wide variety of treatment alternatives, the outcomes are not much different from one another [van Tulder et al. 2000]. Most treatments provide limited contribution to recovery, and some can be worse than no treatment at all [Waddell 1998]. This may be why many patients with LBP do not seek medical attention during acute episodes of LBP [Carey et al. 1996].

Partly in response to the poor treatment outcomes, recent scientific literature has placed an increasing emphasis on encouraging patients to take on responsibility for dealing with their own LBP [Biering-Sørensen and Bendix 2000; Carey et al. 1995]. There is evidence that the following behaviors can provide some assistance for coping with LBP:

- **Remain active.** While limited (<24 hr) bed rest may be indicated in some cases, extended bed rest seems to be harmful to recovery from LBP [Hagen et al. 2005]. Instead, patients are encouraged to maintain as active a lifestyle as possible. There is no indication that staying active is harmful, and staying active is believed to result in improved recovery, more rapid return to work, and less disability [Snook 2004a].

- **Avoid forward bending.** When LBP is present, many patients have difficulty adopting a forward-bending position and such a posture should be avoided [Snook 2004a]. There are indications that avoiding this posture is particularly important during the first hour after waking [Snook et al. 1998]. Aids can be used to help reduce flexion. Examples include chairs with good lumbar support and armrests (which can facilitate standing up), long-handled shoe horns, and reachers to help pick up an object off the floor without bending. Supporting the body with an arm (e.g., against a table or chair) can also minimize bending and decrease the load on the low back [Ferguson et al. 2002]. Lying down on the floor to watch television may be less stressful than sitting in a chair for those with LBP.

- **Exercises.** Most specific back exercises are not recommended during acute episodes of LBP. An exception may be so-called McKenzie back extension exercises, which are passive exercises that extend the spine and require little or no exertion from the back muscles themselves [McKenzie 1985]. The three basic exercises are: (1) lying face down with the head turned to the side, (2) lying down in extension (placing elbows on the ground underneath the shoulders, and (3) extension in lying – perform a pushup, but only raising the upper half of the body (the hips and legs remain on the floor). Studies have shown that these exercises (10 of each up to every 2 hr) can reduce symptoms in 40%–89% of people with LBP [Donelson et al. 1990, 1997].

- **Control the pain.** Nonprescription pain relievers such as aspirin, taken in their proper dosages, seem to be just as effective as other painkillers and have fewer side effects [Snook 2004a]. Neither aspirin nor ibuprofen should be taken if indigestion or ulcers are present. A physician should be consulted if there are any questions or concerns when taking these or other medications.

- **Consult with qualified health providers.** When LBP persists beyond a week or two or worsens despite efforts at control, consultation with a qualified health provider is recommended.

V. SUMMARY AND GUIDELINES

Despite an improvement in back injury rates in mining over the past several years, LBP and low back disability remain costly and significant problems for the U.S. mining industry. Because some factors related to back pain are not under our control, it is not currently possible to eliminate back pain. However, improved job design based on ergonomic principles can prevent an estimated 20%–40%, of work-related back pain, and secondary (postinjury) interventions can reduce lost workdays and related costs by 40%–50% [Snook 2006]. The following are the major points presented in this report.

- While there are several possible causes of LBP, disc degeneration seems to be one of the primary culprits associated with chronic back pain. Disc degeneration seems to be caused by a lack of nutrition to the disc resulting from fractures of the vertebral endplates. These endplates are the first structure to fail in the spine when the spine is placed under load, such as when performing lifting tasks or when exposed to WBV. Section I provides information on the nature of LBP and its causes.

- Experts believe that endplate fractures are probably the result of fatigue failure. Fatigue failure starts when a load (e.g., during lifting) causes a small crack in the endplate. Subsequent loads (e.g., repeated lifting) will cause this crack to expand, leading to a larger fracture.

- Back injury rates have decreased over the past decade; however, these injuries remain the leading cause of disability (lost work time) in mining. Certain mining environments (e.g., coal mines with restricted working heights) impose greater stresses on the back due to the awkward postures (stooping and kneeling) that must be used.

- Research has shown that work in stooping and kneeling postures results in performance limitations, increased spine loading, and increased risk of LBP. Lifting capacity and strength are both reduced in restricted postures, and mobility is greatly reduced. Research suggests that the weight of loads should be decreased by 20%–30% when using restricted postures compared to the acceptable weight in an unrestricted standing position.

- Effective LBP prevention efforts require an integrated and coordinated approach through which problem jobs or tasks are identified and control measures put in place. This approach is known as the risk management cycle. Research indicates that two components are critical to the success of this approach: (1) management commitment (in terms of time and resources), and (2) active worker participation in the process.

- Elimination of manual lifting is one of the best methods for reducing LBP risk. This can be accomplished by streamlining supply-handling systems and by use (or development) of mechanical-assist devices for load transport.

- If materials must be lifted manually, the lifting tasks should be designed to minimize the load experienced by the worker's spine. Three keys to accomplishing this are: (1) reducing the weight of the load, (2) reducing the distance of the load from the body, and (3) limiting the amount that the torso has to bend forward during the lift.

- Psychosocial job stresses, such as job dissatisfaction, monotony, and time stress, have demonstrated clear relationships to LBP incidence rates. An optimal psychosocial work environment for workers is characterized by demands that are adapted to an individual's capacities, a satisfactory level of influence (decision latitude), adequate social support from superiors and colleagues, and a balance between efforts expended at work and reward received.

- Since the factors associated with LBP are not all within our control, it is not realistic to believe that we can currently prevent all cases of LBP. Thus, it is critical to respond appropriately when LBP does occur and do what is necessary to get the injured worker back on the job as soon as practicable (i.e., reduce the disability associated with LBP). Improved workplace design, proactive return-to-work programs, improved communication with the injured worker, and management commitment to the process are important components of a successful disability reduction program.

- There is evidence that certain wellness behaviors can help injured workers cope with episodes of LBP. The following behaviors seem to be of most benefit: (1) stay as active as possible, (2) avoid forward bending of the torso, (3) perform passive back extension exercises, and (4) control the pain with over-the-counter pain medications (if necessary). Consultation with a qualified health provider should be considered if pain worsens or fails to resolve after a couple of weeks.

VI. DIRECTIONS FOR FUTURE RESEARCH

Knowledge of LBP and injury risk in mining has increased greatly over the last two decades. However, much remains to be learned (and perhaps some to be discarded) from what we believe we know presently. The following represents some of the research areas that would be most fruitful, in the author's opinion:

- **Optimizing materials-handling systems at mine sites.** Materials handling is a major cause of back injuries in mining and is a source of many other types of injury as well. However, while there have been isolated assessments of the materials-handling systems at various mine sites, a truly comprehensive evaluation of methods and optimal supply system design has not been undertaken. Such an evaluation could produce a "best practices" document that could have a significant impact on materials-handling practices in the industry. Many benefits could be accrued from such an examination, including decreased injury risk and increased efficiency of operation. All aspects of materials handling should be considered, from packaging and delivery of items by the supplier to final end use.

- **Development of mechanical-assist devices for mining applications.** While there are certainly numerous mine settings where traditional or existing mechanical-assist devices can be effectively used, there are many unique mining applications in need of new mechanical-assist development or modification of existing designs. A particular need from the LBP perspective is in the underground mining environment. As discussed earlier, underground back injury rates are as much as double those at surface operations. This may be a result of the difficulty in providing appropriate mechanical-assist devices in many underground sites. This suggests that efforts should continue to focus on developing and implementing innovative mechanical-assist technologies in underground mines. Furthermore, a compendium of successful applications of effective mechanical-assist designs in mining would be of great benefit to the industry.

- **Coping with work in restricted postures.** Clearly, one of the unique and difficult challenges to preventing back injuries in mining is that of the postural demands placed on workers in low- to midseam coal mines and during the performance of maintenance tasks in both surface and underground mines. Developing improved methods of coping with work in restricted postures should be a priority of LBP prevention efforts. Efforts should focus on minimizing load handling in restricted postures, providing appropriate support mechanisms for the body in these postures, and evaluating schemes for alternating postures to provide appropriate rest periods for body tissues or joints stressed by maintaining a specific posture. Certainly, the development of mechanical-assist devices for low- to midseam mines (as described above) would be an important aspect of such an effort.

- **Improved exposure assessment tools for low back pain risk.** Studies have shown that exposure assessment methods that are quantitative in terms of physical risk factors for LBP do a better job of predicting those jobs or tasks that have a higher risk of LBP. Unfortunately, many available exposure assessment tools used in other industries have limited applicability in the mining environment. New mining-specific exposure assessment tools should be developed to better characterize job risk so that interventions can be better targeted to jobs with high risk. Such assessment tools could leverage new technologies and/or risk models to provide the necessary data for this effort. In particular, improved techniques for the quantitative measurement of "dose" analogous to available measures for noise or chemicals would be of great benefit in characterizing low back risk for specific jobs or tasks in the mining industry.

- **Intervention effectiveness.** Few high-quality scientific intervention studies related to primary (prevention of initial occurrence) and secondary prevention (prevention of future recurrence) of LBP exist in the literature. Most studies currently available consist of retrospective analyses and case reports of intervention effectiveness rather than prospective studies or randomized clinical trials. A clear demonstration of the effectiveness of ergonomics interventions to reduce the rates and costs associated with back injuries would provide persuasive evidence that would be useful in making a business case to mines reluctant to invest in such an approach. Such studies should do more than just document reduction of risk factors or discomfort, but demonstrate real decreases in the occurrence of injuries. Critical to this effort would be development of effective methods of measuring the efficacy and cost-effectiveness of interventions on reducing workplace injuries.

- **Dissemination of information/innovation diffusion studies.** A significant challenge exists in terms of disseminating—to workers, managers, and practitioners—the information necessary to best address the risks of LBP in the mining environment. An improved approach to communications is required to ensure that workers and employers understand and have the information necessary to effectively use the range of workplace interventions available to reduce workplace injury risk. Furthermore, data are needed to determine the effectiveness of dissemination approaches and to ensure that the proper individuals are receiving the information, understand the information, and are able to act effectively with the information provided. Barriers in this process need to be identified and subsequently addressed through rigorous evaluation of information dissemination processes.

REFERENCES

Adams MA, Dolan P [1995]. Recent advances in lumbar spinal mechanics and their clinical significance. Clin Biomech *10*:3–19.

Adams MA, Freeman B, Morrison HP, Nelson IW, Dolan P [2000]. Mechanical initiation of intervertebral disc degeneration. Spine *25*:1625–1636.

Andersson GBJ, Svensson HO, Odén A [1983]. The intensity of work recovery in low back pain. Spine *8*:880–884.

ANSI-HFES (American National Standards Institute-Human Factors and Ergonomics Society) [1988]. Human factors engineering of computer workstations. New York: American National Standards Institute, Inc. ANSI-HFES100-1988.

Armstrong RB [1990]. Initial events in exercise-induced muscular injury. Med Sci Sports Exer *22*:729–735.

Ayoub MM, Bethea NJ, Deivanayagam S, Asfour SS, Bakken GM, Liles P, Mital A, Sherif M [1978]. Determination and modeling of lifting capacity. Final report, grant #5R010H-0054502, HEW, NIOSH. Lubbock, TX: Texas Tech University.

Ayoub MM, Bethea NJ, Bobo M, Burford CL, Caddel DK, Intaranont K, Morrissey S, Selan JL [1981]. Mining in low coal. Vol. I. Biomechanics and work physiology. Lubbock, TX: Texas Tech University. U.S. Bureau of Mines contract No. H0387022. NTIS No. PB 83–258160.

Ayoub MM, Smith, JL, Selan JL, Chen HC, Fernandez JE, Lee YH, Kim HK [1985a]. Manual materials handling in unusual positions: phase II. Final report prepared for the University of Dayton Research Institute.

Ayoub MM, Smith JL, Selan JL, Fernandez JE [1985b]. Manual materials handling in unusual positions: phase I. Final report prepared for the University of Dayton Research Institute.

Basmajian JV, DeLuca C [1985]. Muscles alive: their functions revealed by electromyography. 5th ed. Baltimore, MD: Williams and Wilkins, pp. 260–261.

Bergenudd H, Nilsson B [1988]. Back pain in middle age. Occupational workload and psychologic factors: an epidemiologic survey. Spine *13*:58–60.

Bergqvist-Ullman M, Larsson N [1977]. Acute low back pain in industry. Acta Orthop Scand 170(Suppl):1–117.

Bernard BP, ed. [1997]. Musculoskeletal disorders and workplace factors: a critical review of epidemiologic evidence for work-related musculoskeletal disorders of the neck, upper extremity, and low back. Cincinnati, OH: U.S. Department of Health and Human Services, Public Health Service, Centers for Disease Control and Prevention, National Institute for Occupational Safety and Health, DHHS (NIOSH) Publication No. 97–141.

Biering-Sørensen F, Bendix AF [2000]. Working off low back pain. Lancet *355*:1929–1930.

Bogduk N [1991]. The lumbar disc and low back pain. Neurosurg Clin N Amer *2*:791–806.

Bogduk N [1997]. Clinical anatomy of the lumbar spine and sacrum. 3rd ed. New York: Churchill Livingstone.

Bongers PM, de Winter CR, Kompier MA, Hildebrandt VH [1993]. Psychosocial factors at work and musculoskeletal disease. Scand J Work Environ Health *19*(5):297–312.

Bovenzi M, Hulshof CT [1999]. An updated review of epidemiologic studies on the relationship between exposure to whole-body vibration and low back pain (1986–1997). Int Arch Occup Environ Health *72*(6):351–365.

Brinckmann P, Biggemann M, Hilweg D [1988]. Fatigue fracture of human lumbar vertebrae. Clin Biomech 3(Suppl 1):1–23.

Brinckmann P, Forbin W, Biggemann, M, Tillotson M, Burton K [1998]. Quantification of overload injuries to thoracolumbar vertebrae and discs in persons exposed to heavy physical exertions or vibration at the workplace. Part II: Occurrence and magnitude of overload injury in exposed cohorts. Clin Biomech 13(Suppl 2):1–36.

Burgess-Limerick R, Straker L, Pollock C, Dennis G, Leveritt S, Johnson S [2007]. Participative ergonomics for manual tasks in coal mining. Int J Ind Ergon 37:145–155.

Canyon Research Group [1982]. Human factors design guidelines for personnel carriers. Westlake Village, CA: Canyon Research Group, Inc.

Carayon P, Lim S-Y [2006]. Psychosocial work factors. In: Marras WS, Karwowski W, eds. Interventions, controls, and application in occupational ergonomics. Boca Raton, FL: Taylor and Francis, pp. 5-1 to 5-9.

Carey TS, Garrett J, Jackman A, McLaughlin C, Fryer J, Smucker DR [1995]. The outcomes and costs of care for acute low back pain among patients seen by primary care practitioners, chiropractors, and orthopedic surgeons: the North Carolina back pain project. N Engl J Med 333(14):913–917.

Carey TS, Evans AT, Hadler NM, Lieberman G, Kalsbeek WD, Jackman AM, Fryer JG, McNutt RA [1996]. Acute severe low back pain: a population-based study of prevalence and care-seeking. Spine 21(3):339–344.

Carter JT, Birrell LN, eds. [2000]. Occupational health guidelines for the management of low back pain at work: principal recommendations. London: Faculty of Occupational Medicine.

Cohen AL, Gjessing CC, Fine LJ, Bernard BP, McGlothlin JD [1997]. Elements of ergonomics programs: a primer based on workplace evaluations of musculoskeletal disorders. Cincinnati, OH: U.S. Department of Health and Human Services, Public Health Service, Centers for Disease Control and Prevention, National Institute for Occupational Safety and Health, DHHS (NIOSH) Publication No. 97–117.

Cole DC, Rivilis I, Van Eerd D, Cullen K, Irvin E, Kramer D [2005]. Effectiveness of participatory ergonomic interventions: a systematic review. Toronto, Ontario, Canada: Institute for Work and Health.

Collier SG, Chan WL, Mason S, Pethick AJ [1986]. Ergonomic design handbook for continuous miners. Edinburgh, U.K.: Institute of Occupational Medicine.

Conway, EJ, Unger, RL [1989]. Material handling devices for underground mines. Pittsburgh, PA: U.S. Department of the Interior, Bureau of Mines (BOM) Information Circular 9212.

Cutlip RG [2006]. Soft-tissue pathomechanics. In: Marras WS, Karwowski W, eds. Fundamentals and assessment tools for occupational ergonomics. Boca Raton, FL: Taylor and Francis, pp. 15-1 to 15-43.

Devereux J, Rydstedt L, Kelly V, Weston P, Buckle P [2004]. The role of work stress and psychological factors in the development of musculoskeletal disorders. Norwich, U.K.: Health and Safety Executive Research Report 273.

Deyo RA, Weinstein JN [2001]. Low back pain. N Engl J Med 344:363–370.

Donelson R, Grant W, Kamps C, Medcalf R [1990]. Pain response to sagittal end-range spinal motion: a prospective, randomized, multicentered trial. Spine 16:S206–S212.

Donelson R, Aprill C, Medcalf R, Grant W [1997]. A prospective study of centralization of lumbar and referred pain: a predictor of symptomatic discs and anular competence. Spine 22(10):1115–1122.

Drury CG [1985]. Influence of restricted space on manual materials handling. Ergonomics 28:167–175.

Dul J [1986]. Muscular coordination in working postures. In: Corlett N, Wilson J, Manenica I, eds. The ergonomics of working postures. London: Taylor and Francis, pp. 111–125.

Dul J, Weerdmeester B [1993]. Ergonomics for beginners. London: Taylor and Francis.

Eisenstein AM, Parry CR [1987]. The lumbar facet arthrosis syndrome. J Bone Joint Surg 69B:3–7.

Elkin AJ, Rosch PJ [1990]. Promoting mental health at the work place: the prevention side of stress management. Occup Med 5(4):739–754.

Farfan HF [1973]. Mechanical disorders of the low back. Philadelphia, PA: Lea and Febiger.

Ferguson SA, Marras WS [1997]. A literature review of low back disorder surveillance measures and risk factors. Clin Biomech (Bristol, Avon) 12(4):211–226.

Ferguson SA, Gaudes-MacLaren LL, Marras WS, Waters TR, Davis KG [2002]. Spinal loading when lifting from industrial storage bins. Ergonomics 45:399–414.

Ferguson SA, Marras WS, Burr D [2005]. Workplace design guidelines for asymptomatic vs. low-back-injured workers. Appl Ergon 36:85–95.

Floyd WF, Silver PHS [1955]. The function of the erectores spinae muscles in certain movements and postures in man. J Physiol 129:184–203.

Gallagher S [1991]. Acceptable weights and physiological costs of performing combined manual handling tasks in restricted postures. Ergonomics 34:939–952.

Gallagher S [1997]. Trunk extension strength and muscle activity in standing and kneeling postures. Spine 22(16):1864-1872.

Gallagher S [1999]. Ergonomic issues in mining. In: Karwowski W, Marras WS, eds. The occupational ergonomics handbook. Boca Raton, FL: CRC Press LLC, pp. 1893–1915.

Gallagher S [2002]. Letter to the editor. Spine 27(12):1378–1379.

Gallagher S [2005]. Physical limitations and musculoskeletal complaints associated with work in unusual or restricted postures: a literature review, J Safety Res 36(1):51–61.

Gallagher S, Hamrick CA [1992]. Acceptable workloads for three common mining materials. Ergonomics 35:1013–1031.

Gallagher S, Unger RL [1990]. Lifting in four restricted lifting conditions. Appl Ergon 21:237–245.

Gallagher S, Marras WS, Bobick TG [1988]. Lifting in stooped and kneeling postures: effects on lifting capacity, metabolic costs, and electromyography at eight trunk muscles. Int J Ind Ergon 3:65-76.

Gallagher S, Hamrick CA, Cornelius KM, Redfern MS [2001]. The effects of restricted workspace on lumbar spine loading. Occup Ergon 2:201–213.

Gallagher S, Marras WS, Davis KG, Kovacs K [2002]. Effects of posture on dynamic back loading during a cable lifting task. Ergonomics 45(5):380–398.

Gallagher S, Marras WS, Litsky AS, Burr D [2005]. Torso flexion loads and the fatigue failure of human lumbosacral motion segments. Spine 30(20):2265–2273.

Gallagher S, Marras WS, Litsky AS, Burr D, Landoll J, Matkovic V [2007]. A comparison of fatigue failure responses of old versus middle-aged lumbar motion segments in simulated flexed lifting. Spine *32*(17):1832–1839.

GAO [1997]. Worker protection: private sector ergonomics programs yield positive results. Washington, DC: U.S. General Accounting Office, report GAO/HEHS-97-163.

Garg A, Badger D [1986]. Maximum acceptable weights and maximum voluntary strength for asymmetric lifting. Ergonomics *29*:879–892.

Gaspers K [2002]. Executive decision: workplace safety flourishes when the CEO commits his or her stature. Safety and Health *165*:44–47.

Gibbons LE [1989]. Summary of ergonomics research for the crew chief model development: interim report for period February 1984 to December 1989. Armstrong Aerospace Medical Research Laboratory Report No. AAMRL-TR-90-038. Dayton, OH: Wright-Patterson Air Force Base.

Griffiths DK [1985]. Safety attitudes of management. Ergonomics *28*:61–67.

Hagen KB, Jamtvedt G, Hilde G, Winnem MF [2005]. The updated Cochrane review of bed rest for low back pain and sciatica. Spine *30*:542–546.

Hansson T, Keller T, Spengler D [1987]. Mechanical behavior of the human lumbar spine. II. Fatigue failure during dynamic compressive loading. J Orthop Res *5*:479–508.

Haselgrave CM, Tracy MF, Corlett EN [1997]. Strength capability while kneeling. Ergonomics *34*:939–952.

Hashemi L, Webster BS, Clancy EA, Volinn E [1997]. Length of disability and cost of workers' compensation low back pain claims. J Occup Environ Med *39*:937–945.

Holmström EB, Lindell J, Moritz U [1992]. Low back and neck/shoulder pain in construction workers: occupational workload and psychosocial risk factors. Part 1: Relationship to low back pain. Spine *17*:663–671.

Hoogendoorn WE, van Poppel MNM, Bongers PM, Koes BW, Bouter LM [2000]. Systematic review of psychosocial factors at work and private life as risk factors for back pain. Spine *25*:2114–2125.

Hoogendoorn WE, Bongers PM, de Vet HCW, Houtman ILD, Ariëns GAM, van Mechelen W, Bouter LM [2001]. Psychosocial work characteristics and psychological strain in relation to low-back pain. Scand J Work Environ Health *21*:258–267.

Hunt HA, Habeck RV [1993]. The Michigan disability prevention study: research highlights. Upjohn Institute Staff Working Paper 93-18. Kalamazoo, MI: W. E. Upjohn Institute for Employment Research.

Hunt HA, Habeck RV, VanTol B, Scully SM [1993]. Disability prevention among Michigan employers. Upjohn Institute Technical Report No. 93-004. Kalamazoo, MI: W. E. Upjohn Institute for Employment Research.

ILO (International Labour Organization) [1992]. Conditions of work digest. Vol. 11. Preventing stress at work. Geneva, Switzerland: International Labour Organization.

ISO (International Standards Organization) [1997]. Mechanical vibration and shock: evaluation of human exposure to whole-body vibration. Part 1: General requirements. Geneva, Switzerland: International Organization for Standardization. ISO 2631-1:1997.

Ito M, Incorvaia KM, Yu SF, Fredrickson BE, Yuan HA, Rosenbaum AE [1998]. Predictive signs of discogenic lumbar pain on magnetic resonance imaging with discography correlation. Spine *23*:1252–1260.

Kannel WB, Dawber TR, Kagan A, Revotskie N, Stokes J [1961]. Factors of risk in the development of coronary heart disease--six year follow-up experience. The Framingham study. Ann Intern Med *55*:33–50.

Karasek R, Theorell T [1990]. Healthy work: stress, productivity and the reconstruction of working life. New York: Basic Books.

Kirwan B, Ainsworth LK, eds. [1992]. A guide to task analysis. London: Taylor and Francis.

Klein BP, Jensen RC, Sanderson LM [1984]. Assessment of workers' compensation claims for back strains/sprains. J Occup Med *26*:443–448.

Krause N, Dasinger LK, Neuhauser F [1998]. Modified work and return to work: a review of the literature. J Occup Rehab *8*:113–139.

Kristensen T [1999]. Challenges for research and prevention in relation to work and cardiovascular diseases. Scand J Work Environ Health *25*:550–557.

Kroemer KHE [1997]. Ergonomic design of material handling systems. Boca Raton, FL: Lewis Publishers.

Kulwiec RA, ed. [1985]. Materials handling handbook. New York: John Wiley.

Lavoie-Tremblay M, Bourbonnais R, Viens C, Vézina M, Durand PJ, Rochette L [2005]. Improving the psychosocial work environment. J Adv Nurs *49*(6):655–664.

Lawrence JS [1955]. Rheumatism in coal miners. Part III. Occupational factors. Br J Ind Med *12*:249–261.

Lemstra M, Olszynski WP [2003]. The effectiveness of standard care, early intervention, and occupational management in worker's compensation claims. Spine *28*:299–304.

Lemstra M, Olszynski WP [2004]. The effectiveness of standard care, early intervention, and occupational management in worker's compensation claims: part 2. Spine *29*:1573–1579.

Lindström K [1995]. Finnish research in organizational development and job redesign. In: Murphy LR, Hurrell JJ Jr., Sauter SL, Keita GP, eds. Job stress interventions. Washington, DC: American Psychological Association, pp. 283–293.

Lineberry GT, Adler L [1988]. Mining characterized by encumbered space. Min Sci Tech *6*:125–146.

Linton SJ [2000]. Psychological risk factors for neck and back pain. In: Nachemson AL, Jonsson E, eds. Neck and back pain: the scientific evidence of causes, diagnosis, and treatment. Philadelphia, PA: Lippincott Williams & Wilkins, pp. 57–78.

Loisel P, Abenhaim L, Durand P, Esdaile JM, Suissa S, Gosselin L, Simard R, Turcotte J, Lemaire J [1997]. A population-based, randomized clinical trial on back pain management. Spine *22*(24):2911–2918.

Love AC, Unger RL, Bobick TG, Fowkes RS [1992]. A summary of current Bureau research into the effects of whole-body vibration and shock on operators of underground mobile equipment. Pittsburgh, PA: U.S. Department of the Interior, Bureau of Mines, RI 9439.

Magora A [1973]. Investigation of the relation between low back pain and occupation. IV. Physical requirements: bending, rotation, reaching, and sudden maximal effort. Scand J Rehabil Med *5*:191–196.

Marras WS, Lavender SA, Leurgans SE, Rajulu SL, Allread WG, Fathallah FA, Ferguson SA [1993]. The role of dynamic three-dimensional motion in occupationally-related low back disorders. The effects of workplace factors, trunk position, and trunk motion characteristics on risk of injury. Spine *18*:617–628.

Marras WS, Lavender SA, Leurgans SE, Fathallah FA, Ferguson SA, Allread WG, Rajulu SL [1995]. Biomechanical risk factors for occupationally related low back disorders. Ergonomics *38*:377–410.

Marras WS, Allread WG, Burr DL, Fathallah FA [2000]. Prospective validation of a low-back disorder risk model and assessment of ergonomic interventions associated with manual materials handling tasks. Ergonomics *43*:1866–1886.

Mason S [1992]. Improving the ergonomics of British Coal's mining machinery. Appl Ergon *23*:233–242.

Mayton AG, Merkel R, Gallagher S [1999]. Improved seat reduces jarring/jolting for operators of low-coal shuttle cars. Min Eng *51*(12):52–56.

Mayton AG, Amirouche F, Jobes CC [2005]. Comparison of seat designs for underground mine haulage vehicles using the absorbed power and ISO 2631–1(1985)-based ACGIH threshold limit methods. Int J Heavy Vehicle Syst *12*(3):225–238.

Mayton AG, Jobes CC, Kittusamy NK, Ambrose DH [2006]. Field evaluation of seat designs for underground coal mine shuttle cars. Pittsburgh, PA: U.S. Department of Health and Human Services, Public Health Service, Centers for Disease Control and Prevention, National Institute for Occupational Safety and Health, DHHS (NIOSH) Publication No. 2007–100, IC 9493.

McKenzie R [1985]. Treat your own back. 5th ed. Wellington, New Zealand: Spinal Publications.

McPhee B [1993]. Ergonomics for the control of sprains and strains in mining. National Occupational Health and Safety Commission (Worksafe Australia).

Mishra DK, Fridén J, Schmitz MC, Lieber RL [1995]. Anti-inflammatory medication after muscle injury. A treatment resulting in short-term improvement but subsequent loss of muscle function. J Bone Joint Surg Am *77*:1510–1519.

Mitterer D [1999]. Back injuries in EMS. Emerg Med Serv *28*(3)41–48.

Moneta GB, Videman T, Kaivanto K, Aprill C, Spivey M, Vanharanta H, Sachs BL, Guyer RD, Hochsculer SH, Raschbaum RF, Mooney V [1994]. Reported pain during lumbar discography as a function of anular ruptures and disc degeneration: a reanalysis of 833 discograms. Spine *19*:1968–1974.

Moore SM, Bauer ER, Steiner LJ [2007]. Prevalence and cost of cumulative injuries over two decades of technological advances: a look at underground coal mining in the U.S. SME preprint 07–067. Littleton, CO: Society for Mining, Metallurgy, and Exploration, Inc.

Nachemson AL [2000]. Introduction. In: Nachemson AF, Jonsson E, eds. Neck and back pain: the scientific evidence of causes, diagnosis, and treatment. Philadelphia, PA: Lippincott Williams & Wilkins.

Nachemson AL, Vingård E. [2000]. Assessment of patients with neck and back pain: a best-evidence synthesis. In: Nachemson AF, Jonsson E, eds. Neck and back pain: the scientific evidence of causes, diagnosis, and treatment. Philadelphia, PA: Lippincott Williams & Wilkins.

NIOSH [2006]. Technology news 521 – Development of a mobile manipulator to reduce lifting accidents. Spokane, WA: U.S. Department of Health and Human Services, Public Health Service, Centers for Disease Control and Prevention, National Institute for Occupational Safety and Health, DHHS (NIOSH) Publication No. 2007–111.

NIOSH, Cal/OSHA, CNA Insurance Companies, and Material Handling Industry of America [2007]. Ergonomic guidelines for manual material handling. Cincinnati, OH: U.S. Department of Health and Human Services, Public Health Service, Centers for Disease Control and Prevention, National Institute for Occupational Safety and Health, DHHS (NIOSH) Publication No. 2007–131.

NRC (National Research Council) [2001]. Musculoskeletal disorders and the workplace: low back and upper extremities. Washington, DC: National Academy Press.

NSC (National Safety Council) [2004]. Injury facts, 2004 edition. Itasca, IL: National Safety Council.

OSHA [1990]. Ergonomics program management guidelines for meatpacking plants. Washington, DC: U.S. Department of Labor, Occupational Safety and Health Administration, OSHA report No. 3123.

Pheasant S [1986]. Bodyspace: anthropometry, ergonomics and design. London: Taylor and Francis.

Punnett L, Fine LJ, Keyserling WM, Herrin GD, Chaffin DB [1991]. Back disorders and nonneutral trunk postures of automobile assembly workers. Scand J Work Environ Health *17*:337–346.

Remington PJ, Andersen DA, Alakel MN [1984]. Assessment of whole body vibration levels of coal miners. Volume II: Whole body vibration exposure of underground coal mining machine operators. Cambridge, MA: Bolt Beranek and Newman, Inc. U.S. Bureau of Mines contract No. J0308045. NTIS No. PB 87-144119.

Rivilis I, Cole DC, Frazer MB, Kerr MS, Wells RP, Ibrahim S [2006]. Evaluation of a participatory ergonomic intervention aimed at improving musculoskeletal health. Am J Ind Med *49*:801–810.

Sanders MS, Peay JM [1988]. Human factors in mining. Pittsburgh, PA: U.S. Department of the Interior, Bureau of Mines, IC 9182. NTIS No. PB 88-232467.

Sauter, SL, Hurrell, JJ, Cooper CL, eds. [1989]. Job control and worker health. Chichester, U.K.: John Wiley & Sons.

Schwarzer AC, Aprill CN, Derby R, Fortin J, Kine G, Bogduk N [1994]. Clinical features of patients with pain stemming from the lumbar zygapophysial joints. Is the lumbar facet syndrome a clinical entity? Spine *19*:1132–1137.

Schwarzer AC, Aprill CN, Bogduk N [1995a]. The sacroiliac joint in chronic low back pain. Spine *20*:31–37.

Schwarzer AC, Aprill CN, Derby R, Bogduk N, Kine G [1995b]. The prevalence and clinical features of internal disc disruption in patients with chronic low back pain. Spine *20*:1878–1883.

Seeley PA, Marklin RW [2003]. Business case for implementing two ergonomic interventions at an electric power utility. Appl Ergon *34*(5):429–439.

Selan J [1997]. Mining. Presented at the joint NIOSH/OSHA conference, "Ergonomics: Effective Workplace Practices and Programs," (Chicago, IL, January 8–9, 1997). [http://www.cdc.gov/niosh/topics/ergonomics/EWconf97/ec4selan.html]. Date accessed: May 2008.

Sestito JP, Lunsford RA, Hamilton AC, Rosa RR, eds. [2004]. Worker health chartbook, 2004. Cincinnati, OH: U.S. Department of Health and Human Services, Public Health Service, Centers for Disease Control and Prevention, National Institute for Occupational Safety and Health, DHHS (NIOSH) Publication No. 2004–146.

Siegrist J [1996]. Adverse health effects of high-effort/low-reward conditions. J Occup Psych *1*:27–41.

Snook SH [2004a]. Self-care guidelines for the management of nonspecific low back pain. J Occup Rehab *14*:243–253.

Snook SH [2004b]. Work-related low back pain: secondary intervention. J Electromy Kines *14*:153–160.

Snook SH [2006]. Secondary intervention for low back pain. In: Karwowski W, Marras WS, eds. The occupational ergonomics handbook. 2nd ed. Interventions, controls, and applications in occupational ergonomics. Boca Raton, FL: Taylor and Francis, pp. 24-1 to 24-15.

Snook SH, Webster BS, McGorry RW, Fogleman MT, McCann KB [1998]. The reduction of chronic, non-specific low back pain through the control of early morning lumbar flexion: a randomized controlled trial. Spine *23*:2601–2607.

Solomonow M [2004]. Ligaments: a source of work-related musculoskeletal disorders. J Electromy Kines *14*:49–60.

Solomonow M, Zhou BH, Harri, M, Lu Y, Baratta RV [1998]. The ligamento-muscular stabilizing system of the spine. Spine *23*:2552–2562.

Staal JB, Hlobil H, van Tulder MW, Köke AJA, Smid T, van Mechelen W [2002]. Return-to-work interventions for low back pain: a descriptive review of contents and concepts of working mechanisms. Sports Med *32*:251–267.

Svensson H, Andersson GBJ [1983]. Low-back pain in 40- to 47-year old men: work history and work environment factors. Spine *8*(3):272–276.

Torma-Krajewski J, Lehman M [2007]. Ergonomic interventions at Badger Mining Corporation [2007]. SME preprint 07–066. Littleton, CO: Society for Mining, Metallurgy, and Exploration, Inc.

Torma-Krajewski J, Steiner LJ, Lewis P, Gust P, Johnson K [2006]. Ergonomics and mining: charting a path to a safer workplace. Pittsburgh, PA: U.S. Department of Health and Human Services, Public Health Service, Centers for Disease Control and Prevention, National Institute for Occupational Safety and Health, DHHS (NIOSH) Publication No. 2006–141, IC 9491.

Torma-Krajewski J, Hipes C, Steiner LJ, Burgess-Limerick RJ [2007]. Ergonomic interventions at Vulcan Materials Company. SME preprint 07–065. Littleton, CO: Society for Mining, Metallurgy, and Exploration, Inc.

van Tulder MW, Waddell G [2000]. Conservative treatment of acute and subacute low back pain. In: Nachemson A, Jonsson E, eds. Neck and back pain: the scientific evidence of causes, diagnosis, and treatment. Philadelphia, PA: Lippincott Williams & Wilkins.

Vernon-Roberts B, Pirie CJ [1977]. Degenerative changes in the intervertebral discs of the lumbar spine and their sequelae. Rheumatol Rehabil *16*(1):13–21.

Videman T, Battié MC [1996]. A critical review of the epidemiology of idiopathic back pain. In: Weinstein JN, Gordon SL, eds. Low back pain: a scientific and clinical overview. Rosemont, IL: American Academy of Orthopedic Surgeons.

Waddell G [1998]. The back pain revolution. Edinburgh, U.K.: Churchill Livingstone.

Waddell G, Burton AK [2000]. Occupational health guidelines for the management of low back pain at work: evidence review. London: Faculty of Occupational Medicine.

Wasiak R, Pransky GS, Webster BS [2003]. Methodological challenges in studying recurrence of low back pain. J Occup Rehab *13*:21–31.

Wassell JT, Gardner LI, Landsittel DP, Johnston JJ, Johnston JM [2000]. A prospective study of back belts for prevention of back pain and injury. JAMA *284*(21):2727–2732.

Waters TR, Putz-Anderson V, Garg A, Fine LJ [1993]. Revised NIOSH equation for the design and evaluation of manual lifting tasks. Ergonomics *36*(7):749–776.

Waters TR, Putz-Anderson V, Garg A [1994]. Applications manual for the revised NIOSH lifting equation. Cincinnati, OH: U.S. Department of Health and Human Services, Public Health Service, Centers for Disease Control and Prevention, National Institute for Occupational Safety and Health, DHHS (NIOSH) Publication No. 94–110.

Waters TR, Baron SL, Piacitelli LA, Putz-Anderson V, Skov T, Haring-Sweeney M, Wall DK, Fine LJ [1999]. Evaluation of the revised NIOSH lifting equation: a cross-sectional epidemiologic study. Spine *24*(4):386–395.

Waters TR, Dick RB, Davis-Barkley J, Krieg EF [2007]. A cross-sectional study of risk factors for musculoskeletal symptoms in the workplace using data from the general social survey (GSS). J Occup Environ Med *49*(2):172–184.

Westrin CG [1970]. Low back sick-listing. A nosological and medical insurance investigation. Acta Sociomed Scand *2*(2):127–134.

Winn FJ, Biersner RJ [1992]. Exposure probabilities to ergonomic hazards among miners. Poster presented at the 36th Annual Meeting of the Human Factors Society (Atlanta, GA, October 12–16, 1992).

Zurada J, Karwowski W, Marras W [2004]. Classification of jobs with risk of low back disorders by applying data mining techniques. Occup Ergon *4*:291–305.

APPENDIX.—TABLES USED FOR CALCULATION OF THE REVISED NIOSH LIFTING EQUATION

Table A-1.—Frequency multiplier table

Frequency Lifts/min (F)	Work Duration					
	≤ 1 hr		> 1 but ≤ 2 hr		> 2 but ≤ 8 hr	
	V < 30	V ≥ 30	V < 30	V ≥ 30	V < 30	V ≥ 30
0.2	1.00	1.00	.95	.95	.85	.85
0.5	.97	.97	.92	.92	.81	.81
1	.94	.94	.88	.88	.75	.75
2	.91	.91	.84	.84	.65	.65
3	.88	.88	.79	.79	.55	.55
4	.84	.84	.72	.72	.45	.45
5	.80	.80	.60	.60	.35	.35
6	.75	.75	.50	.50	.27	.27
7	.70	.70	.42	.42	.22	.22
8	.60	.60	.35	.35	.18	.18
9	.52	.52	.30	.30	.00	.15
10	.45	.45	.26	.26	.00	.13
11	.41	.41	.00	.23	.00	.00
12	.37	.37	.00	.21	.00	.00
13	.00	.34	.00	.00	.00	.00
14	.00	.31	.00	.00	.00	.00
15	.00	.28	.00	.00	.00	.00
>15	.00	.00	.00	.00	.00	.00

Table A-2.—Coupling multiplier table

Coupling Types	Coupling multiplier	
	V ≥ 30 in (75 cm)	V ≥ 30 in (75 cm)
Good	1.00	1.00
Fair	0.95	1.00
Poor	0.90	0.90

www.ingramcontent.com/pod-product-compliance
Lightning Source LLC
Chambersburg PA
CBHW081839170526
45167CB00007B/2844